Alcohol and Human Health

Introducing Health Sciences: A Case Study Approach

Series editor: Basiro Davey

Seven case studies on major topics in global public health are the subject of this multidisciplinary series of books, each with its own animations, videos and learning activities on DVD. They focus on: access to clean water in an overcrowded and polluted world; the integration of psychological and biological approaches to pain; alcohol consumption and its effects on the body; the science, risks and benefits of mammography screening for early breast cancer; chronic lung disease due to smoke pollution – a forgotten cause of millions of deaths worldwide; traffic-related injuries, tissue repair and recovery; and the causes and consequences of visual impairment in developed and developing countries. Each topic integrates biology, chemistry, physics and psychology with health statistics and social studies to illuminate the causes of disease and disability, their impacts on individuals and societies and the science underlying common treatments. These case studies will be of value to anyone who is, or wants to be, working in a health-related occupation where scientific knowledge could enhance your prospects. If you have a wide-ranging interest in human sciences and want to learn more about global health issues and statistics, how the body works and the scientific rationale for screening procedures and treatments, this series is for you.

Titles in this series

Water and Health in an Overcrowded World, edited by Tim Halliday and Basiro Davey

Pain, edited by Frederick Toates

Alcohol and Human Health, edited by Lesley Smart

Screening for Breast Cancer, edited by Elizabeth Parvin

Chronic Obstructive Pulmonary Disease: A Forgotten Killer, edited by Carol Midgley

Trauma, Repair and Recovery, edited by James Phillips

Visual Impairment: A Global View, edited by Heather McLannahan

Alcohol and Human Health

Edited by Lesley Smart

Published by Oxford University Press, Great Clarendon Street, Oxford OX2 6DP in association with
The Open University, Walton Hall, Milton Keynes MK7 6AA.

OXFORD
UNIVERSITY PRESS

Oxford University Press is a department of the University of Oxford. It furthers the University's
objective of excellence in research, scholarship, and education by publishing worldwide in

Oxford New York

Auckland Cape Town Dar es Salaam Hong Kong Karachi Kuala Lumpur Madrid Melbourne
Mexico City Nairobi New Delhi Shanghai Taipei Toronto

with offices in

Argentina Austria Brazil Chile Czech Republic France Greece Guatemala Hungary
Italy Japan Poland Portugal Singapore South Korea Switzerland
Thailand Turkey Ukraine Vietnam

Oxford is a registered trade mark of Oxford University Press in the UK and in certain
other countries.

Published in the United States by Oxford University Press Inc., New York

First published 2007

Edited and designed by The Open University.

Typeset by SR Nova Pvt. Ltd, Bangalore, India.

Printed and bound in the United Kingdom at the University Press, Cambridge.

This book forms part of the Open University course SDK125 *Introducing Health Sciences: A Case
Study Approach*. Details of this and other Open University courses can be obtained from the Student
Registration and Enquiry Service, The Open University, PO Box 197, Milton Keynes MK7 6BJ,
United Kingdom:
tel. +44 (0)870 333 4340, email general-enquiries@open.ac.uk.

http://www.open.ac.uk

British Library Cataloguing in Publication Data available on request

Library of Congress Cataloging in Publication Data available on request

ISBN 9780 1992 3735 7

10 9 8 7 6 5 4 3 2 1

ABOUT THIS BOOK

This book and the accompanying material on DVD present the third case study in a series of seven, under the collective title *Introducing Health Sciences: A Case Study Approach*. Together they form an Open University (OU) course for students beginning the first year of an undergraduate programme in Health Sciences. Each case study has also been designed to 'stand alone' for readers studying it in isolation from the rest of the course, either as part of an educational programme at another institution, or for general interest and self-directed study.

The *Alcohol and Human Health* case study is for anyone seeking a scientific understanding of the effects that drinking alcohol has on their body, on individual behaviour and on wider society. The case study starts by looking at current global trends in drinking and its social consequences. It goes on to cover the alcohol journey through the body – the chemical reactions it undergoes as it is broken down, the number of calories released, the damage to the liver, and the effect on the brain and behaviour. Possible treatments for alcoholism are investigated and finally we look at the contested evidence for the positive effects of drinking. On the DVD we use computer animations and videos to underpin the basic science of alcohol at the molecular, psychological and physiological levels, explore the attitudes of young people to drinking, and listen to accounts of people recovering from alcoholism.

No previous experience of studying science has been assumed and new concepts and specialist terminology are explained with examples and illustrations. There is some mathematical content: the emphasis is mainly on interpreting data in tables and graphs, but the text also introduces you step-by-step to some ways of performing calculations that are commonly used in science.

To help you plan your study of this material, we have included a number of 'icons' in the margin to indicate different types of activity which have been included to help you develop and practise particular skills. This icon ⊗ indicates when to undertake an activity on the accompanying DVD. You will need to 'run' the DVD programs on your computer because they are *interactive,* and this function doesn't operate on a domestic DVD-player. The DVD has several computer animations: we illustrate the structures and bonding of some molecules, particularly ethanol (the 'alcohol' molecule), the diffusion of ethanol into the bloodstream and the organs, and signalling mechanisms in the brain. The social effects of drinking are illustrated by interviews with international university students, and with three recovering alcoholics and a senior administrator of Alcoholics Anonymous.

Activities involving pencil-and-paper exercises are indicated by this icon 🖎 , and if you need a calculator you will see 🖩. Some additional activities for Open University students only are described in a *Companion* text, which is not available outside the OU course. These are indicated by this icon 📖 in the margin. Some activities involve using the internet and are marked by this icon 🌐. References to activities for OU students are given in the margins of the book and should not interrupt your concentration if you are not studying it as part of an OU course.

At various points in the book, you will find 'boxed' material of two types: Explanation Boxes and Enrichment Boxes. The Explanation Boxes contain basic concepts explained in the kind of detail that someone who is completely new to the health sciences is likely to want. The Enrichment Boxes contain extension material, included for added interest, particularly if you already have some knowledge of basic science. If you are studying this book as part of an OU course, you should note that the Explanation Boxes contain material that is *essential* to your learning and which therefore may be *assessed*. However, the content of the Enrichment Boxes will *not* be tested in the course assessments.

The authors' intention is to bring you into the subject, develop confidence through activities and guidance, and provide a stepping stone into further study. The most important terms appear in **bold** font in the text at the point where they are first defined, and these terms are also in bold in the index at the end of the book. Understanding of the meaning and uses of the bold terms is essential (i.e. assessable) if you are an OU student.

Active engagement with the material throughout this book is encouraged by numerous 'in text' questions, indicated by a diamond symbol (◆), followed immediately by our suggested answers. It is good practice always to cover the answer and attempt your own response to the question before reading ours. At the end of each chapter, there is a summary of the key points and a list of the main learning outcomes, followed by self-assessment questions to enable you to test your own learning. The answers to these questions are at the back of the book. The great majority of the learning outcomes should be achievable by anyone who has studied this book and its DVD material; one or two learning outcomes for some chapters are only achievable by OU students who have completed the *Companion* activities, and these are clearly identified.

Internet database (ROUTES)

A large amount of valuable information is available via the internet. To help OU students and other readers of books in this series to access good quality sites without having to search for hours, the OU has developed a collection of internet resources on a searchable database called ROUTES. All websites included in the database are selected by academic staff or subject-specialist librarians. The content of each website is evaluated to ensure that it is accurate, well presented and regularly updated. A description is included for each of the resources.

The website address for ROUTES is: http://routes.open.ac.uk/

Entering the Open University course code 'SDK125' in the search box will retrieve all the resources that have been recommended for this book. Alternatively if you want to search for any resources on a particular subject, type in the words which best describe the subject you are interested in (for example, 'alcohol'), or browse the alphabetical list of subjects.

Authors' acknowledgements

As ever in The Open University, this book and DVD combine the efforts of many people with specialist skills and knowledge in different disciplines. The principal authors were Basiro Davey (public health), James Phillips (biology), Frederick Toates (psychology), Lesley Smart (chemistry), Sotiris Missailidis (biochemistry) and Tom Heller (health and social care). Our contributions have been shaped and immeasurably enriched by the OU course team who helped us to plan the content and made numerous comments and suggestions for improvements as the material progressed through several drafts. It would be impossible to thank everyone personally, but we would like to acknowledge the help and support of academic colleagues who have contributed to this book (in alphabetical order of discipline): Nicolette Habgood, Heather McLannahan, Carol Midgley (biology), Jeanne Katz (health and social care), Elizabeth Parvin (physics) and Peter Naish (psychology). The media developers who contributed directly to the production of the DVD animations were Steve Best, Greg Black, Eleanor Crabb and Brian Richardson. Audiovisual material was developed by Hendrik Ball (independent producer) and Jo Mack (OU Sound and Vision) and by Basiro Davey, James Phillips, Lesley Smart and Frederick Toates.

We are very grateful to our External Assessor, Professor Susan Standring, Head of Department of Anatomy and Human Sciences, Kings College London, whose detailed comments have contributed to the structure and content of the book and kept the needs of our intended readership to the fore.

Special thanks are due to all those involved in the OU production process, chief among them Joy Wilson and Dawn Partner, our wonderful Course Manager and Course Team Assistant, whose commitment, efficiency and unflagging good humour were at the heart of the endeavour. We also warmly acknowledge the contributions of our editor, Bina Sharma, whose skill has improved every aspect of this book; Steve Best, our graphic artist, who developed and drew all the diagrams; Sarah Hofton, our graphic designer, who devised the page designs and layouts; and Martin Keeling, who carried out picture research and rights clearance. The activity to support Open University students in developing information literacy skills was devised by Clari Hunt (OU Library). The media project managers were Judith Pickering and James Davies.

For the copublication process, we would especially like to thank Jonathan Crowe of Oxford University Press and, from within The Open University, Christianne Bailey (Media Developer, Copublishing). As is the custom, any small errors or shortcomings that have slipped in (despite our collective best efforts) remain the responsibility of the authors. We would be pleased to receive feedback on the book (favourable or otherwise). Please write to the address below.

Dr Basiro Davey, SDK125 Course Team Chair

Department of Biological Sciences
The Open University
Walton Hall
Milton Keynes
MK7 6AA
United Kingdom

Environmental statement

Paper and board used in this publication is FSC certified.

Forestry Stewardship Council (FSC) is an independent certification, which certifies that the virgin pulp used to make the paper/board comes from traceable and sustainable sources from well-managed forests.

CONTENTS

The DVD activities associated with this book were written, designed and developed by Hendrik Ball, Steve Best, Greg Black, Eleanor Crabb, Basiro Davey, Jo Mack, James Phillips, Brian Richardson, Lesley Smart and Frederick Toates.

CHAPTER 1

ALCOHOL AND ITS EFFECTS ON HEALTH

1.1 Alcohol has a long history

Humans have drunk alcohol for at least 12 000 years and it has been used in religious rituals in ancient cultures as diverse as Samaria, Babylon, Egypt, the Chinese Imperial court and Anglo-Saxon Britain (Figure 1.1). The ancient Romans had a god of wine (Bacchus); so did the ancient Greeks (Dionysus). Christian Communion services and certain Jewish religious rituals include wine to the present day. Alcohol has more than ritual significance: wine was routinely drunk in Mediterranean countries and, further north, beer in particular was part of the staple diet until the early 20th century, and was probably a safer drink than the often-contaminated water of earlier times – food for the body as well as a blessing from the gods. When people raise a glass of alcohol to 'toast' each other, they often reflect this benevolent view: the English say 'Good health!' or 'Cheers!'; the French say 'À votre santé!' (to your health); and in Germany 'Pröst!' (may it do you good).

Yet attitudes to alcohol vary greatly around the world. In many nations, it is an accepted way to 'unwind' from the pressures of life, a common accompaniment to meals and many social occasions. Home-brewed beer and distilled spirits are drunk throughout Africa and South America. By contrast, in Islamic and Buddhist cultures alcohol is generally prohibited. Throughout its history, drinking alcohol to excess has been associated with deviant behaviour and harm, as another term for drunkenness – *intoxication*, from the Latin *toxicum*, a poison (as in *toxic*) – signifies. In a few ancient cultures the ability to drink huge quantities of alcohol was considered a sign of masculinity, for example, among

(a)

(b)

Figure 1.1   (a) Egyptian wall painting: picking grapes; from the tomb of Nakht, the Royal Astronomer and keeper of the Royal Vineyards, 18th dynasty (c. 1567–1320 BC) (Photo: Held Collection/The Bridgman Art Library). (b) Carving depicting a man putting a tap in a barrel; from a choir stall in the Abbey of St Lucien in Beauvais, France (15th century AD). (Photo: Musée National du Moyen Age et des Thermes de Cluny, Paris, Lauros/Giraudon/ The Bridgman Art Library)

followers of Dionysus. Echoes of this attitude can be found on Saturday nights among young men in some western city centres.

The advocacy of total **abstinence** from alcohol began as early as AD 200, but it is best known from the 'temperance' movement in predominantly Christian countries in the 19th century, which gave rise to the term 'teetotaller' (short for 'temperance total') – someone who deliberately abstained from alcohol. Alcohol was banned in the USA during the 'prohibition era' from 1920 to 1933. Mississippi was the last state to repeal its prohibition laws in 1966. In 2000, around 4 billion abstainers worldwide outnumbered alcohol drinkers by about two to one, but the ratio is shifting rapidly as alcohol drinking spreads into countries with little previous use and women take up the habit in increasing numbers.

In this chapter, we look at global data on how much alcohol is drunk in different parts of the world and relate this to the impact of alcohol on health. The World Health Organization (WHO, 2002) estimated that in the year 2000, alcohol-related diseases, accidents, violence and suicide caused 1.8 million deaths worldwide. Next, we turn to the basic chemistry of the *ethanol* molecule – the chemical name for alcohol (Chapter 2), and consider how it is broken down in the body and what waste products are produced. Molecular models on the DVD are used to demonstrate how the structure of ethanol underlies its effects on the body. In Chapters 3 and 5, with the help of DVD animations, we describe how alcohol is absorbed into the bloodstream and where it goes, and in the process we answer the question 'what causes a hangover?' In Chapter 4 we explore the effects of alcohol on behaviour, motor control and memory, and look at the causes of addiction and the possibilities for treatment. Finally, we consider the evidence that small amounts of alcohol may be beneficial to health – at least for some people (Chapter 6).

1.2   Problems in estimating alcohol consumption

Knowing how much alcohol people actually drink is essential in evaluating the harm or benefit it may cause. But estimating alcohol consumption accurately is fraught with difficulties.

◆ Think of some reasons why this is so.

◆ 1   The concentration (strength) of alcohol varies between types of drink and few people know the difference. Does a glass of red wine contain more or less alcohol than a vodka shot?

2   The standard volume of a 'drink' varies between types of drink and between countries. How does the alcohol content of an English pint of real ale compare with a half-litre of German lager, an American 'ounce' or a British 'optic' of spirits?

3   Different ways of referring to the amount of alcohol in a drink make comparison difficult. For example, wine, beer and spirit labels usually

give the alcohol content as a *percentage* of the liquid in the bottle. 'Drink safely' advice generally refers to *units* of alcohol often quoted in terms of 'a glass of wine' or a 'half-pint of beer' or a 'measure of spirits'. (How to calculate these quantities is covered in Activity 1.1 and Vignette 1.2.)

Asking people to estimate how much alcohol they drink in (say) a typical week produces very inaccurate results. When drinks are poured at home, the 'measure' is often uncontrolled and estimates tend to be even more inaccurate (Figure 1.2). Population surveys consistently find that drinkers' own estimates of the total amount consumed account for only 40–60% of the alcohol *sold* (Bloomfield et al., 2003). People who drink a lot of alcohol may say they drink less when asked, because excessive drinking is disapproved of in most societies. Conversely, some individuals may exaggerate their drinking because it carries high status among their peers.

Now try Activity 1.1.

Figure 1.2 These are the standard glasses of wine sold in the UK, containing 125 ml, 175 ml and 250 ml. For the red wine being poured (which contains 14% pure alcohol) they contain 1.75, 2.45 and 3.5 UK units respectively – far more than most people suspect. (Photo: Lesley Smart)

Activity 1.1 Calculating the volume of pure alcohol in drinks

Allow 10 minutes

This activity will give you practice in doing some simple calculations involving percentages. You can use a calculator if you find this helpful.

The international system of scientific units of measurement (SI) uses a set of seven base units to derive all others. The ones most relevant in this case study are length (*metre*, m), mass (*kilogram*, kg) and time (*second*, s). In SI units, a litre should properly be referred to as one decimetre cubed, 1 dm³ (the volume contained in a cube of side-length $\frac{1}{10}$ m), but in this book the more familiar **litre** will be used as the unit for measuring the volume of liquids; one litre is written as 1 l.

A litre can be divided into 100 **centilitres** (100 cl) or 1000 **millilitres** (1000 ml), so there are 10 ml in every centilitre. A standard bottle of wine contains three-quarters of a litre of liquid, which is usually expressed on the label in centilitres as 75 cl. Most other liquid products are usually labelled in litres or in millilitres (ml).

How do you know how much pure alcohol is in a 75 cl bottle of wine? The label always gives the percentage of the total volume that is pure alcohol – as in 12% vol. (it typically varies between 12 and 15% vol.). If 12% of a 75 cl bottle of wine is pure alcohol, what is the volume of pure alcohol in the bottle?

Percentages are always fractions of 100, so 12% could be written as 12/100 or 'twelve hundredths'. The question posed above about the wine bottle could have been stated as 'What is twelve hundredths of 75?'

The total volume of pure alcohol in a 75 cl bottle of wine labelled '12% vol.' is:

$$75 \text{ cl} \times \frac{12}{100} = 9 \text{ cl}$$

If you are more used to thinking in 'imperial' weights and measures, one litre is the metric equivalent of 1.76 pints.

If you convert the volume in the bottle from cl to ml, the calculation would be:

$$750 \text{ ml} \times \frac{12}{100} = 90 \text{ ml}$$

◆ What volume of pure alcohol would you consume if you drank half a litre of beer with an alcohol content of 5%? Express your answer in millilitres (ml).

◆ Half a litre is 500 ml. If 5% of 500 ml is pure alcohol, the beer contains:

$$500 \text{ ml} \times \frac{5}{100} = 25 \text{ ml of pure alcohol}$$

When trying to estimate alcohol consumption, data based on alcohol *sales* alone ignore the substantial amounts brought in from outside a country. In 2001, this 'under the counter' importation amounted to 2 litres of pure alcohol (the equivalent of more than 20 bottles of wine) for every adult in the UK (Leifman, 2001). In parts of Africa and South America, up to a quarter of the alcohol consumed in rural areas is locally produced and not counted in official sales figures. The estimates of alcohol consumption in Section 1.3 include all legal and illegal sources.

1.3 Variations in alcohol consumption around the world

Alcohol consumption varies hugely between countries with different cultural traditions. To enable comparison, the amount of alcohol consumed in each population must be converted into *litres of pure alcohol* (Figure 1.3). At one end of the spectrum are Muslim countries such as Pakistan, which in 2000 averaged 20 ml of pure alcohol consumed per person (0.02 litres or about 4 teaspoons); at the opposite extreme are many European countries which averaged over 13 litres per person per year. Note that country-based averages disguise the fact that some people abstain completely, some drink 'little but often' and others regularly get drunk. The amount consumed by each *individual* is what determines the effect on their health.

The average amount of pure alcohol consumed *globally* has remained at about 5 litres per person per year since the 1970s. But this apparent stability hides an overall *decline* in consumption in 'developed' countries (taken together) to around 9 litres per person by the year 2000, and an *increase* to about 3 litres per person on average in 'developing' countries (as a group). (Definitions of these categories are given in Box 1.1 on page 6.) The WHO is particularly concerned about the rising trend in developing countries, where there is little public awareness of the potential risks, and services to prevent or treat alcohol-related harm are scarce. Table 1.1 shows four developing countries with the fastest rising alcohol consumption. Between them they account for over 40% of the world's population.

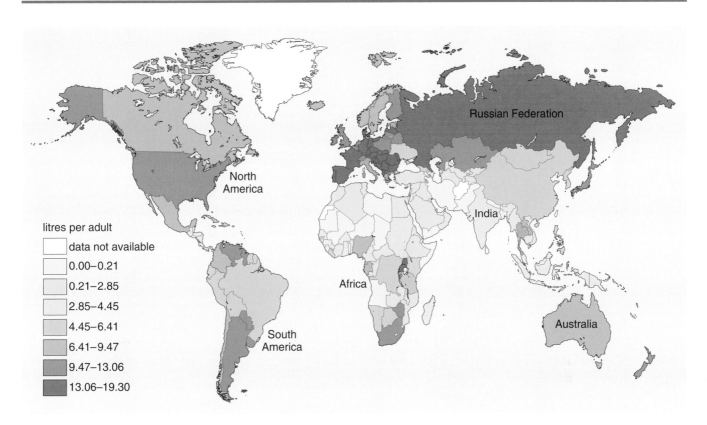

Figure 1.3 World map showing variations in average consumption in litres of pure alcohol per person aged 15 years and over in the year 2000 or 2001. (Source: data derived from WHO, 2005b, Figure 1, p. 1; additional data from the *WHO Global Alcohol Database* online)

Table 1.1 Average alcohol consumption in litres of pure alcohol per person aged 15 years and over in China, India, Brazil and North Korea, 1970–2001. (Source: data from WHO, 2006)

Country	1970 (litres)	1980 (litres)	1990 (litres)	2001 (litres)	2005 population (thousands)*
China	0.8	1.7	3.3	4.5	1 315 844
India	0.4	0.5	0.7	0.8	1 103 371
Brazil	2.8	3.5	4.3	5.3	186 405
North Korea§	2.6	2.6	3.7	5.7	47 817
					total 2 653 437

*Population data are often stated in 'millions' or 'thousands' of people, as in the right-hand column of Table 1.1. Thus, in 2005 China had 1 315 844 'thousands' of people, or 1 315 844 000, which is just over 1.3 billion (a billion is one thousand million). §Democratic People's Republic of Korea.

Box 1.1 (Explanation) Defining developed and developing countries

The terms 'developed' and 'developing' as applied to countries are problematic: different definitions are used by different international agencies, using criteria such as gross national income and/or development indicators such as the proportion of children attending school or access to clean drinking water. **Developed countries** include all those in Western Europe and North America, plus Australia, New Zealand, Japan and some others; they have high national incomes and high scores on an index of development indicators. Those with lower national incomes and development scores are the **developing countries**, but the variation within this group is huge, ranging from the world's poorest nations (e.g. Sierra Leone, Burkina Faso) to rapidly emerging economic powers such as India and China. In some classifications, Eastern European countries such as the Russian Federation, Poland and the Ukraine are referred to as 'transitional economies'.

If you are studying this book as part of an Open University course, go to Activity C1 in the *Companion* now.

If the changes over time in the total *amounts* of alcohol in Table 1.1 look small to you, consider the *rate* at which they have been rising and the number of people affected. The amount consumed per person *doubled* in India in the period shown and in China it increased more than five times. As largely rural cultures in Asia, Africa and South America have been exposed to rapid urbanisation in recent decades, the increased level of stress associated with life in the 'human zoo' – from overcrowding, poverty, pollution, unemployment, violence and crime – has gone hand-in-hand with the increasing consumption of alcohol. The damaging effects of alcohol seem set to become another 'export' from richer to poorer nations.

The differences in alcohol consumption between countries – even those with similar levels of wealth and industrial development – can be striking, as Table 1.2 illustrates. Study the table, then turn to Activity 1.2.

Table 1.2 Average alcohol consumption in litres of pure alcohol per person aged 15 years and over, in selected high-income countries between 1970 and 2000. (Source: data from WHO, 2006)

Notes of consumption patterns

it rose to 1980, then fell below 1970 level

rose throughout period; little change after 1990

Country	1970	1980	1990	2000
Australia	11.5	13.0	10.5	9.2
Finland	6.0	8.1	9.9	10.0
France	23.2	20.1	16.7	13.4
Ireland	7.0	9.6	11.2	14.2
Italy	21.2	17.9	11.7	9.3
Japan	4.5	5.9	6.9	7.5
Spain	16.1	18.6	13.4	11.9
UK	8.5	10.8	10.8	10.2
USA	9.4	10.7	9.5	8.4

Activity 1.2   Seeing patterns in Table 1.2

Allow about 10 minutes

Look along the top two rows of Table 1.2 and note the general trend in alcohol consumption in Australia and in Finland over time. In the space to the right of the table, we have summarised the patterns in a few words. Now do the same for the other rows in the table. Then answer the following questions.

◆ Which countries in Table 1.2 have shown a rising trend and which a falling trend in alcohol consumption in this period?

◆ People in countries with relatively low alcohol consumption in 1970 (Finland, Ireland and Japan) have tended to drink a *bit more* in each successive decade, whereas populations with the highest alcohol consumption at the start of the period (France, Italy and Spain) drank a *lot less* over time. Countries with 'middling' values in 1970 (Australia, UK and USA) didn't change as much in 30 years.

By 2000, the differences in alcohol consumption between these 'developed' countries were not as great as they had been in 1970. They were tending towards the average for Europe of around 13 litres of pure alcohol per adult per year.

Comment

If you want to find out how alcohol consumption has been changing in any other countries, you can consult the WHO's *Global Alcohol Database* (see Further reading at the end of this book).

However, the *pattern* of alcohol consumption is even more important than the total *amount* consumed. Drinking large amounts in one go – particularly of spirits – has a much higher health risk than drinking the same quantity of alcohol in a more dilute form (e.g. beer or wine) in smaller amounts spread over several days. Figure 1.4 (overleaf) shows the same map as in Figure 1.3, but this time the colours indicate the levels of *risk* associated with differences in the typical drinking patterns in different parts of the world. Spend a few minutes comparing the two maps.

If you are studying this book as part of an Open University course, go to Activity C2 in the *Companion* now.

◆ Compare the average consumption of alcohol in South America shown in Figure 1.3 with the risk level shown in Figure 1.4. What strikes you about the comparison?

◆ Populations in the northern part of the continent (e.g. in Brazil) tend to drink a *lot less* alcohol on average than those in the south (Argentina in particular), but their drinking pattern carries a much *higher risk.*

Several other parts of the world show this surprising reversal (e.g. compare India and Africa in the two maps), which is largely due to **binge drinking** ('drinking to get drunk'). Typically, binge drinkers consume very little alcohol on most days of the week, but get extremely drunk at weekends or on festive occasions. The rapid rise to a high level of alcohol in the bloodstream is particularly dangerous, as you

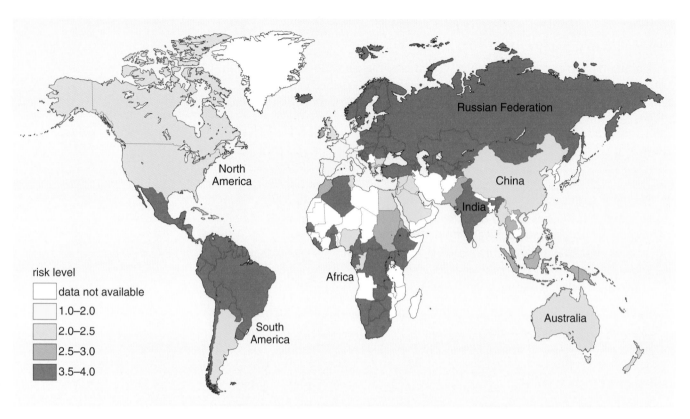

Figure 1.4 Differences in alcohol drinking patterns in different regions of the world, ranging from level 1, 'least risky' (drinking with meals, mainly wine) to level 4, 'most risky' (frequent heavy drinking to become intoxicated, mainly beer and spirits). (Source: data derived from WHO, 2005b, Figure 2, p. 2)

will see later in this chapter. Binge drinking is increasing among young people in most of Europe, particularly young women – as Rachael's story illustrates.

1.4 Alcohol, youth and gender – Rachael's story

In most countries where alcohol is consumed regularly, men drink far more than women and are much more likely to become intoxicated. This is particularly true in most developing countries, where women rarely drink at all. But in some developed countries, including the UK, the gender difference began narrowing from the late 1980s. By 2000, women in Western Europe had the highest average alcohol consumption of females anywhere in the world (WHO, 2005b). We will illustrate the effect of alcohol through a series of fictional 'vignettes' about a young woman called Rachael, based on real people's stories. Rachael is introduced in Vignette 1.1. We do not claim that 'Rachael' is representative even of women in Britain, let alone of women elsewhere. It is important to remember the specific culture and geographical region in which the vignette is situated and recognise that there will be similarities and differences with people in other circumstances or from elsewhere in the world.

Patterns of drinking may be established at quite a young age and may be a response to individual features in each young person's situation. Table 1.3 shows some personal and environmental factors that might influence alcohol misuse amongst adolescents.

Vignette 1.1   Rachael's early drinking experiences

Rachael grew up in a rural area of England. She was born in 1974 into a stable family with a reasonable income. Her mother worked as a classroom assistant in a local school, her dad was a maintenance engineer and she had three older brothers. Both parents enjoyed a drink and most of their social evenings revolved around the pub in a nearby village. Rachael can remember playing in the family room of the pub as a child and liking the friendly chatter of people in the bar. As an adolescent it seemed natural for her to experiment with alcohol, and by her early teens she was drinking every weekend. This behaviour wasn't unusual in England in her age-group, as Figure 1.5 shows.

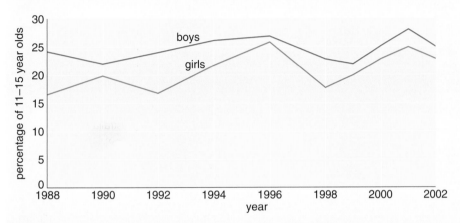

Figure 1.5 Percentage of 11–15 year olds in England who reported having an alcoholic drink in the preceding week, 1988–2002. (Source: data derived from Boreham and Shaw, 2002)

◆   What do you notice about the trends over time in Figure 1.5?

◆   As time passes, the graph lines get closer together because the percentage of girls who said they drank alcohol was rising slightly faster than it was for boys.

Table 1.3 Possible risk factors for alcohol and other substance misuse in adolescents. (Source: data derived from Bonomo and Proimos, 2005)

Temperament and personality traits – Behaviour and moods that meet the diagnostic criteria for 'antisocial personality disorder' or 'sensation-seeking trait'.

Emotional and behavioural problems – Conduct disorder, depression, attention deficit hyperactivity disorder (ADHD).

Familial factors – Family attitudes that favour the use of alcohol, particularly if parents abuse it; poor or inconsistent parenting.

Early onset of alcohol use – Drinking alcohol before the age of 15 increases the risk of alcohol dependence by about four times compared with drinkers who delay until young adulthood.

Poor social connections – At school and in the community.

Peer-group members use alcohol – A strong predictor of alcohol misuse in young people.

In the UK the quantity of alcohol consumed in a single drinking session is measured in 'units', where one unit is the equivalent of 10 ml of pure alcohol regardless of the type of drink involved. The size of a standard 'unit' of pure alcohol varies hugely between countries (see Table 1.4). Note that the UK has the *smallest* unit at 10 ml, just over half that of the USA. Table 1.5 lists the number of (UK) units in some commonly available drinks. So how much did Rachael drink on a Saturday night (see Vignette 1.2)?

Table 1.4 Amount of pure alcohol in the standard 'unit' of various countries (Source: Tapson, 2004)

Country	ml	Country	ml
Australia	12.7	Italy	12.7
Canada	17.1	Japan	25.0
Denmark	15.2	Netherlands	12.5
Finland	13.9	New Zealand	12.7
France	15.2	Portugal	17.7
Hungary	21.5	Spain	12.7
Iceland	12.0	UK	10.0
Ireland	12.7	USA	17.7

Table 1.5 Number of (UK) units of alcohol in common alcoholic drinks.

A pint of ordinary strength lager (Carling Black Label, Fosters)	2 units
A pint of strong lager (Stella Artois, Kronenbourg 1664)	3 units
A pint of ordinary bitter (John Smith's, Boddingtons)	2 units
A pint of best bitter (Fuller's ESB, Young's Special)	3 units
A pint of ordinary strength cider (Woodpecker)	2 units
A pint of strong cider (Dry Blackthorn, Strongbow)	3 units
A 175 ml glass of red or white wine	around 2 units
A pub measure 'optic' of spirits	1 unit
An alcopop (e.g. Smirnoff Ice, Bacardi Breezer, WKD, Reef)	around 1.5 units

Vignette 1.2   Rachael and 'binge drinking'

When Rachael went out with her friends as a teenager, she typically drank three bottles of 'alcopops' (fruit-flavoured drinks containing a spirit such as vodka, 1.5 UK units each), followed by a 'large glass' of white wine (250 ml, around 3 units) – and a tequila 'slammer' (2 units).

◆ In Rachael's mind she had only had five 'drinks' but how many units of alcohol had she drunk on a typical night out? How much is this in ml of pure alcohol?

◆ $(3 \times 1.5) + 3 + 2 = 9.5$ units, or 95 ml of pure alcohol in a single session. (This is 75 ml more than the average amount consumed per person in Pakistan in a whole year.)

The amount of alcohol that constitutes 'binge drinking' varies between countries, but in the UK it is defined as 8 or more units in a single drinking session for a man and 6 or more units for a woman (you will see why shortly). On this basis, Rachael has a serious drink problem, but she is not alone. Figure 1.6 shows that drinking to get drunk is common among 15-year-olds in many European countries, and is particularly high in the UK.

In 2005, a report from the WHO summed up the problem that Rachael typifies:

'…drinking to excess among the general population and heavy episodic drinking among young people are on the rise in many countries throughout the world. The reasons may be the increased availability of alcoholic beverages, aggressive marketing and promotion of such drinks aimed at young people, and a breakdown in the lines of authority and taboos related to age. Young drinkers in developing countries are increasingly emulating the drinking styles that are identified as those of the developed world.'

(WHO, 2005a, paragraph 9, p. 2)

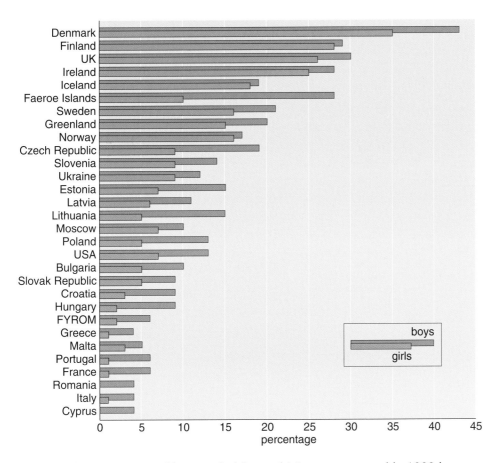

Figure 1.6 Percentage of boys and girls aged 15 years surveyed in 1999 in a range of European countries (and Moscow and the USA for comparison) who say they have been drunk 10 or more times during the last 12 months. (FYROM is the Former Yugoslav Republic of Macedonia.) (Source: data adapted from Hibell et al., 2000)

1.5   Alcohol and the world's health

'Alcohol is the anaesthesia by which we endure the operation of life.'

(George Bernard Shaw, 1856–1950)

Shaw's observation has an uncomfortable ring of truth in most westernised cultures. Most interviewees in a UK survey (see Figure 1.7) said they drank because alcohol made them feel happy and relaxed, but they also used it to *self-medicate*, i.e. combat depression and anxiety, forget problems and boost confidence. The rising proportion of people in non-Muslim developing countries who are now also using alcohol may do so in part to ease the stress of urban living.

Of course, the consequences of drinking alcohol aren't always harmful. Most people who drink it enjoy the taste and the pleasurable effects on mood and sensation (discussed in Chapter 4). An unusual feature of alcohol **epidemiology** (i.e. the statistical study of its association with health outcomes in different countries and population groups) is that in small doses it appears to do no harm to most people and for some it may have health benefits (Chapter 6). The WHO (2002) estimates that if everyone in 'developed' countries stopped drinking alcohol, there would be about 17% more cases of *ischaemic stroke* (damage to part of the brain caused by lack of oxygen and nutrients). Similarly, the risk of developing ischaemic heart disease, due to blocked coronary arteries, may also be reduced by drinking small amounts of alcohol regularly. However, in most countries, 'deaths prevented' by drinking alcohol are hugely outweighed by 'deaths caused'.

Ischaemia (iss-kee-mee-ah) means inadequate blood supply due to an obstructed blood vessel.

We noted earlier that 1.8 million people died in 2000 from alcohol-related causes – just over 3% of all deaths worldwide in that year (WHO, 2002). Alcohol ranked *fifth* in the list of global risks to health and caused 4% of the global burden of disease, disability and premature death (as measured in DALYs, see Box 1.2). Averaging these statistics across the world's six billion inhabitants disguises the disproportionate impact in *developed* countries, where alcohol ranked *third* among risks to health and contributed 9.2% of all DALYs. Of even greater concern is the

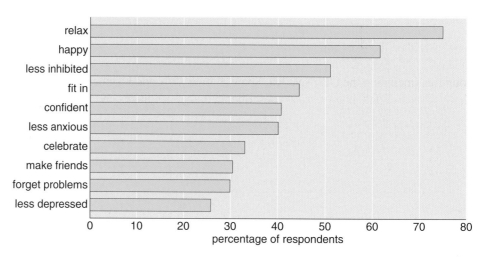

Figure 1.7   Results of a UK survey of 1000 people asked for their reasons for drinking alcohol. (Data derived from Mental Health Foundation, 2006, Figure 1, p. 13)

fact that alcohol was the *largest* cause of disease, disability and premature death in developed countries among people aged 15–29 years.

> **Box 1.2** (Explanation) Disability Adjusted Life Years (DALYs)
>
> The WHO estimates the extent of ill health in a population due to a disease, disorder or disability (e.g. excessive alcohol consumption) in units known as **DALYs** ('dailies' – **Disability Adjusted Life Years**), which aim to reflect the real impact of each disease, disorder or disability on people's lives. The calculation is complex, but in essence DALYs combine an estimate of the number of years lived with a reduced quality of life, taking into account the severity of the condition (every condition is assigned a 'weighting factor' to reflect this), *and* the number of years of life lost if the person dies prematurely, based on their age and the average life expectancy in that population.
>
> If the total number of disability adjusted life years suffered by all the people affected by a condition in a particular country, in a given year, are added together, some very large numbers result.

1.5.1 Alcohol and acute personal harm

Alcohol has both 'acute' and 'chronic' effects. **Acute effects** are the *short-term* consequences of becoming intoxicated; the symptoms escalate quite rapidly as more alcohol is consumed, but most people appear to make a relatively quick recovery. However, a significant minority of drunken episodes result in injury, alcoholic poisoning or loss of life. Vignette 1.3 (overleaf) illustrates how Rachael's drinking behaviour exposed her to several risks.

Rachael is a fictional character but in Activity 1.3 you will be able to compare the accounts we have created of her behaviour with real-life experiences.

Activity 1.3 Students' attitudes to alcohol

Allow about 30 minutes

Now would be the ideal time to study the video 'Students' attitudes to alcohol' on the DVD associated with this book. In the video, university students from several countries (including the UK, Sweden, Spain, Latvia, Italy and Denmark) – all in their first year at an English university – describe their drinking habits and the effects on their mood, social interactions and behaviour, and their attitudes to risk-taking. They refer to the age when they began drinking alcohol, the amounts they drink and their attitudes to alcohol and health. They were interviewed by two of the authors of this book. As you watch the video, pay particular attention to the following questions:

One of the students refers to the risk of 'STIs', which is an abbreviation of 'sexually transmitted infections'.

- Do they know how much they drink on a 'big night out'?
- Are they concerned about the possible effects of binge-drinking on their health?

We will ask you to think back to other aspects of these interviews in Chapter 4 of this book. If you can't study the video now, continue with the chapter and complete the activity as soon as you can.

Vignette 1.3   Minor accidents and taking risks

Although Rachael continued to drink quite heavily at weekends, she completed her school exams with good enough grades for a place on a university degree course. She supplemented her income by getting an evening job in a pub, and on Saturdays she often went 'clubbing' with friends, generally coming home drunk in a taxi or someone's car. Most Sundays were spent recovering from a hangover and she rarely made it to Monday morning lectures, but mostly she kept up with her coursework. Once she fell down stairs and broke her wrist, but apart from bumps and bruises while drunk she suffered no other physical injuries. However, twice she had sex with men she hardly knew when she was too drunk to consider the consequences, and was anxious afterwards that she might be pregnant or have caught a sexually transmitted infection, but everything was okay. Rachael considered herself to have been lucky. A male student in her year choked to death on his own vomit as he lay drunk, unconscious and alone. When she graduated with a lower second-class degree, Rachael got a job in the human resources department of a major UK travel agent.

◆ What harms other than those in Rachael's story may follow swiftly after excessive drinking?

◆ Traffic accidents, street crime, anti-social behaviour, violence (Figure 1.8), the use of illegal drugs, suicide attempts, accidents in the workplace and truancy from school are all associated with excessive alcohol consumption.

Figure 1.8   There are around 1.2 million incidents of alcohol-related violence in the UK every year. (Photo: Matt Cardy/Getty Images).

The students in DVD Activity 1.3 refer to incidents where people were hurt as a result of excessive alcohol consumption. Table 1.6 summarises some findings of a report on alcohol-related injuries and violence in the UK.

Table 1.6 Some indicators of alcohol-related injuries and violence in the UK. (Source: data derived from Academy of Medical Sciences, 2004, and Alcohol Concern Briefings and Fact Sheets, 2005)

In a typical year in the UK:

- Over 1 million people and one-third of attendances at Accident and Emergency departments are alcohol-related.

- Around 150 000 hospital in-patient admissions (40% of the total) are due to alcohol-related conditions.

- Over 17 000 people are injured in traffic accidents involving alcohol, including 3000 serious injuries and around 500 fatalities; the number of injuries has been falling, but the proportion of *fatal* accidents due to alcohol has been rising since the early 1990s.

- Around one-third of the 360 000 annual incidents of domestic violence are linked to alcohol misuse.

- There are 1.2 million incidents of alcohol-related violence; 47% of the victims of violent crime interviewed for the British Crime Survey believed their attacker had been drinking.

- Alcohol may be a factor in up to 65% of suicide attempts; alcohol is frequently consumed prior to suicide even by individuals who *don't* have a drinking problem.

Gender, alcohol, accidents and violence

Gender plays a large part in the epidemiology of alcohol. The majority of those affected by alcohol-related injuries and violence are men, particularly in developed countries. Figure 1.9 shows the 'alcohol attributable risks' for a number of major types of accident and injury in males and females, i.e. the lengths of the bars show what proportion of different types of accidents and of injuries can be attributed to alcohol.

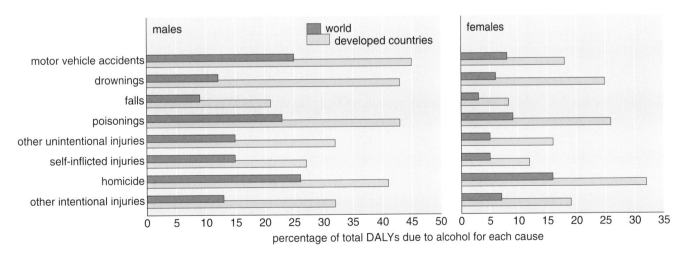

Figure 1.9 Percentage of disability and premature death (measured in DALYs) due to accidents and intentional injuries, worldwide and for developed countries, which can be attributed to alcohol: data for males and females in 2000. (Source: data derived from WHO, 2004, Table 20, p. 55)

◆ Figure 1.9 shows that more than 30% of the DALYs sustained in some types of accident and violence can be attributed to alcohol. Which are they, where do they occur and which sex is most affected?

◆ More than 30% of disability and premature death among *males* in *developed* countries as a result of motor vehicle accidents, drowning, poisonings, and other unintentional and intentional injuries, are attributed to alcohol; so are more than 30% of homicides in *both* sexes in developed countries.

Acute alcoholic poisoning in the Russian Federation

The experience of Russia since the break-up of the Soviet Union in the early 1990s illustrates the potential for a large intake of alcohol to cause death very quickly. Russia is unusual in that the majority of alcohol-related deaths are among men in their 40s and 50s and are due to acute **alcoholic poisoning** – intoxication so extreme that it leads to unconsciousness that can result in death, particularly when drinkers collapse out of doors in low temperatures (Pridemore and Kim, 2006). By 2001, alcohol consumption had risen to amongst the highest in the world. Nearly a third of all deaths in Russia are believed to be directly or indirectly the result of excessive alcohol consumption (Nemtsov, 2005). A fall in average male life expectancy (to 59 years in 2001) has been attributed primarily to the effects of 'binge' drinking of distilled spirits such as vodka, which comprise around 80% of the alcohol consumed. Alcohol is also associated with a range of chronic health and social problems in Russia, as in all countries where excessive drinking has become commonplace.

1.5.2 Alcohol, gender and chronic health problems

The prolonged excessive consumption of alcohol is the principal cause or a major contributory factor in over 60 diseases (WHO, 2004). The long-term or **chronic effects** on individuals who regularly abuse alcohol, develop gradually over many years and are largely *irreversible*. They include **alcoholic liver cirrhosis** (sir-oh-sis), which occurs when the liver is so damaged by prolonged excessive consumption of alcohol that scar tissue replaces a large proportion of its normal structure (as we describe in Chapter 5). This severely impairs the liver's ability to remove toxic substances (including alcohol) from the blood

Table 1.7 gives the DALYs in 2000 due to alcohol-related chronic diseases. Figure 1.10 shows the extent to which some important chronic conditions are attributed to alcohol in developed countries and worldwide. More recent data from a huge European study added colon and rectum cancer to the list (Ferrari et al., 2007).

Although in every country a larger *number* of males suffer chronic health problems due to excessive alcohol, it is not apparent from Table 1.7 and Figure 1.10 that women seem to be more *vulnerable* to alcohol-related diseases, particularly alcoholic liver cirrhosis and brain damage. They are more likely to

Homicide refers to the killing of one person by another; it combines murder and lesser charges (e.g. manslaughter in the UK) which vary between countries.

Table 1.7 Worldwide alcohol-related burden of chronic disease in 2000, expressed in thousands of disability adjusted life years (DALYs) due to alcohol. (Source: data derived from WHO, 2004, Table 16, p. 51)

Disease/disorder	DALYs (thousands)		
	Male	**Female**	**Total**
alcohol-related birth disorders (mainly *fetal alcohol syndrome*) (Section 5.5)	68	55	123
alcohol-related cancers (mainly of the mouth, larynx, oesophagus, stomach, intestines, rectum, liver and possibly some breast cancers)	3180	1021	4201
cardiovascular disease (heart disease, strokes, high blood pressure) due to alcohol (Chapter 5)	4411	(−428)	3983
neuropsychiatric conditions due to alcohol (e.g. alcohol dependence, alcoholic psychosis, epilepsy and depression) (Chapter 4)	18 090	3814	21 904
other chronic diseases due to alcohol (mainly alcoholic cirrhosis of the liver and diabetes) (Section 5.3.1)	3695	860	4555

The negative value for alcohol-related cardiovascular disease among women indicates that the number of DALYs *prevented* by drinking alcohol exceeded the number *caused* by alcohol; the 'net profit' was 428 thousand DALYs.

develop a serious illness at a *lower* level of alcohol consumption than men and women who drink heavily suffer a higher mortality rate than males who drink equivalent amounts. A Danish study of around 13 000 men and women followed for 12 years, demonstrated that the level of drinking above which there was a significant risk of liver disease was 7 to 13 'beverages' per week in women, but 14 to 27 'beverages' per week in men (Becker et al., 1996).

A 'beverage' was defined in this study as a Danish unit of alcohol, 15.2 ml, see Table 1.4.

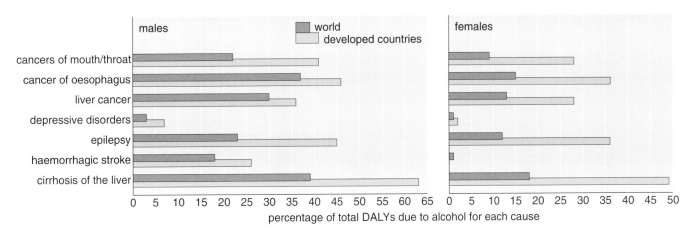

Figure 1.10 Percentage of disability and premature death (measured in DALYs) due to major chronic diseases, worldwide and for developed countries, which can be attributed to alcohol: data for males and females in 2000. (Source: as Figure 1.9)

An explanation of how comparable data were generated for Figure 1.11 from countries with different population structures is given in Box 1.3. The reasons for these gender differences are discussed in Chapter 5.

Alcoholic cirrhosis of the liver in Great Britain

Figure 1.11 presents trends in total deaths from liver cirrhosis, comparing England, Wales and Scotland with the average for 12 Western European countries.

◆ What do you notice about the patterns of liver cirrhosis in the countries shown in Figure 1.11?

◆ Deaths from liver cirrhosis in other European countries began declining in the 1970s and fell below Scottish rates somewhere between 1990 and 2000. This downturn did not occur in Britain, where rates continued rising in both sexes and both age-groups. The death rate in Scotland was greater than in England and Wales throughout the period.

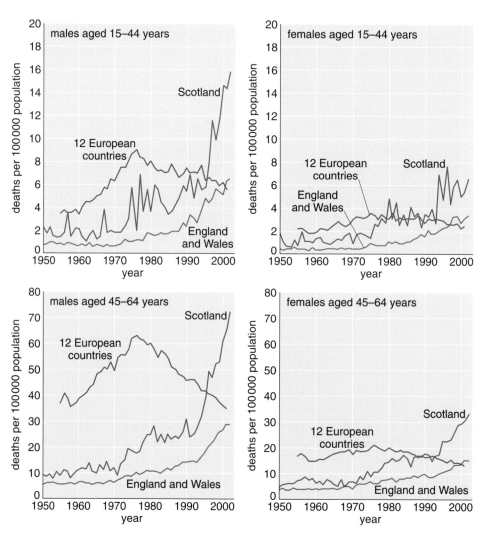

Figure 1.11   Trends in deaths from liver cirrhosis per 100 000 population in England, Wales and Scotland, compared with the combined average for 12 Western European countries, in two age-groups and for males and females separately, between 1950 and 2002. (Note that the upper and lower graphs use different *scales* on the vertical axes; mortality rates were age-standardised using the European standard population; see Box 1.3.) (Source: data derived from Leon and McCambridge, 2006)

The death rates in Figure 1.11 were higher for men than for women of the same age in all countries throughout the period, but the general 'shape' of the graph lines for women resemble those for men 10–20 years earlier. This suggests that cirrhosis in women may catch up with male rates in the future. Now look at the two different age-groups. The shape of the graph lines is remarkably similar, but in both sexes the older group were dying at several times the rate of their younger counterparts.

Box 1.3 (Explanation)   Age-standardised data and countries with different age structures

Epidemiological data from different countries are often 'age-standardised'. There are two reasons. First, the proportion of young, middle-aged and older people differs between populations, i.e. they have different **population age-structures**. In particular, the proportion of young people is higher in most developing countries, where birth rates are relatively high and life expectancy is not as great as in more affluent nations. Conversely, developed countries have 'ageing' populations, with a much higher proportion of older people and fewer in younger age-groups.

Secondly, many diseases and causes of death affect people in certain age-groups more than in others: for example, traffic injuries are more likely among children and young adults, whereas alcoholic liver cirrhosis mostly affects older people.

◆ Average alcohol consumption in England and Nigeria was similar in 2000. Explain why comparing the *number* of deaths per 100 000 population from liver cirrhosis in England with the *number* per 100 000 in Nigeria might be misleading.

◆ You would expect the numbers dying (per 100 000 population) to be higher in England than in Nigeria just because England has a higher proportion of older people in its population who are most at risk from liver cirrhosis. This makes direct comparison of rates in the two populations difficult.

The solution is to perform a mathematical adjustment, called **age-standardisation**, which takes account of differences in population age-structures between the countries whose data you want to compare; the method involves taking a 'reference population' (in Figure 1.11 it is the whole of Europe) and using its population age-structure as the standard for adjusting the populations of interest (e.g. England and Nigeria) so that their health data can be directly compared.

1.5.3 Harm to the families of heavy drinkers

Diseases and disorders due to alcohol are only part of the catalogue of chronic harm. The families of heavy drinkers also suffer major long-term effects, for example from domestic violence, which shows a strong association with alcohol use in countries all over the world (WHO, 2004). For example, Table 1.8 overleaf presents responses to interviews in 2400 households in Nepal, but similar results could have come from any country where alcohol is a growing problem.

Table 1.8 Sources of alcohol-related harm in 2400 Nepalese families. (Note that some people report more than one problem.) (Source: data derived from Dhital et al., 2001)

Percentage of adults reporting that their children suffered as a result of alcohol in the family:	%
violence and physical abuse	33.4
neglect and mental abuse	28.5
deprivation from education	20.2
encouraged children to use intoxicants	11.1

Percentage of children reporting that parental drinking had a negative impact in their family:	%
domestic violence	40.0
loss of wealth and indebtedness	27.8

In the UK, 30–60% of child protection cases involve alcohol and up to 1.3 million children may be adversely affected by alcohol abuse in their family (Academy of Medical Sciences, 2004). Children of problem drinkers have higher than average levels of behavioural and emotional disturbances (Figure 1.12); they experience more difficulties at school and are themselves more likely to develop alcohol and drug-related problems later in life. Marriages where one partner drinks heavily are twice as likely to end in divorce. This is a growing threat in Rachael's adult life (Vignette 1.4).

Figure 1.12 Children and young people who grow up in families where a parent frequently gets drunk suffer an increased risk of depression and other emotional problems. (Photo: Bryan Rosengrant/ Flickr Photo Sharing)

Vignette 1.4 Heavy drinking affects Rachael's work and family

After Rachael had been in her job at the travel company for several years she became a middle manager. When she was 28 she married a joiner called Andy and they had a child, Jamie, now aged four. Rachael was able to drink heavily at weekends and still maintain her domestic and working life, but she concealed how much she drank. She was becoming **alcohol tolerant**, i.e. she had to drink much more than in the past to achieve the same effect. In the office she was seen as the 'life and soul' of any party, but occasionally after a 'binge' she was unable to get to work and pretended to be ill. Sometimes she took risks, including driving while her blood-alcohol was above the legal limit, and drinking more than was safe while looking after Jamie. Each New Year's Eve she resolved to control her drinking, but always resumed her usual pattern. Andy found it hard to discuss the problem with Rachael, because it always ended in a terrible row and she tended to get drunk afterwards. He found ways to shield Jamie from seeing his mother drinking and gradually became the primary carer for their son. Andy increasingly resented the cost of alcohol in the family budget, which amounted to over £60 in an average week.

1.6 Economic costs of alcohol-related harm

Rachael is not alone in spending heavily on alcohol. Its manufacture, distribution, advertising and sales are major contributors to the economies of virtually all developed nations. Alcohol is big business: for example, £4.6 billion was spent on alcohol in London alone during the year 2000 (NERA, 2003). The marketing policies of the major alcohol suppliers are being increasingly criticised for targeting younger people, where long-term profit will follow if they become regular drinkers. Estimates in the USA show that the drinks industry spends at least US$120 million a year advertising alcoholic drinks in magazines for adolescents (Garfield et al., 2003) and US$28 million placing alcohol adverts in the most popular TV programmes amongst teenagers. The cash value of *underage* drinking in the USA in 2001 has been estimated at US$22.5 billion (Foster et al., 2006).

Governments try to balance the maintenance of income from taxing the production and sale of alcoholic drinks, with preservation of the individual's freedom to drink alcohol and the protection of the public. Figure 1.13 comes from an appraisal by the British government of the personal and societal costs and benefits of alcohol, but this framework can be applied to any country in the world. The total annual cost to the British economy was estimated in 2004 at about £30 billion per annum, or £500 per head of population (Cabinet Office Strategy Unit, 2004). This included £12–18 billion in working days lost and £1.4–1.7 billion in health care costs. Studies from other developed, and a number of developing countries, tell a similar story; for example, 10–30% of absenteeism from work in India, Costa Rica and Bolivia is directly due to the effects of alcohol (WHO, 2004).

◆ Look at Figure 1.13 and think back to the interviews with university students which you watched on the DVD during Activity 1.3. Which events in their accounts would have involved financial costs to society as a whole?

◆ They recounted incidents of injuries to people who were drunk, or who had been injured in drunken fights or dangerous stunts; these would

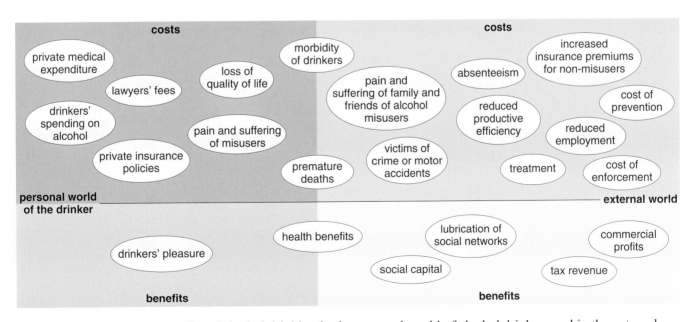

Figure 1.13   Costs and benefits of alcohol drinking in the personal world of alcohol drinkers and in the external world in which they live. (Source: adapted from Cabinet Office, 2003)

have resulted in substantial health-care costs. In the most serious cases of disorder or violence, there would have been costs in police time and possibly also criminal court proceedings. The student whose jaw was broken missed a lot of time studying; if he had to repeat a term, this would have cost the state, and possibly his parents, additional tuition costs and living expenses.

Later in this book, we explain how alcohol interacts with the major organs, body systems and mental processes to produce its 'self-medicating' effects and cause physical and psychological harm. But first you need to know what alcohol *is*. The chemistry of alcohol is the subject of Chapter 2.

If you are studying this book as part of an Open University course, go to Activity C3 in the *Companion* now.

Summary of Chapter 1

1.1 Alcohol has been drunk for thousands of years, and is associated with health, pleasure and religious rituals in some cultures, but is considered sinful, damaging and is prohibited in others.

1.2 Estimating the amount of alcohol consumed is difficult because the alcohol content and the 'measures' of standard drinks vary, people underestimate and misrepresent their level of consumption, and 'duty free', illegal imports and home-made alcohol may be omitted.

1.3 Average alcohol consumption varies from under 20 ml of pure alcohol per person per year in Muslim and Buddhist countries, to over 13 litres annually per European; consumption is falling in most developed countries but rising in most non-Muslim developing countries.

1.4 Worldwide, males are more likely than females to drink regularly and to 'binge drink', but in the UK, female alcohol consumption is catching up with male rates; young people in developing countries increasingly emulate this drinking pattern.

1.5 Alcohol in small amounts can have health benefits in some individuals, but regularly exceeding this level can result in short and long-term risks to health.

1.6 Over 1.8 million people died in 2000 and over 58 million DALYs were lost as a result of alcohol-related injuries, poisoning and chronic diseases.

1.7 Men are far more likely than women to suffer alcohol-related harm, particularly in developed countries, because they drink more; alcoholic poisoning is particularly common among Russian men. Females are more *vulnerable* to chronic illness at a lower level of alcohol consumption than their male counterparts.

1.8 Alcohol-related injuries and violence are more prevalent among younger age-groups and chronic diseases are commoner at older ages; alcoholic liver cirrhosis is rising particularly quickly in Great Britain and cases are occurring at younger ages.

1.9 Excessive alcohol consumption damages families through marital breakdown, domestic violence, loss of employment and disrupted education among children.

1.10 The alcohol industry spends large sums on marketing and has been accused of targeting its advertising at young people.

1.11 Countries with significant levels of alcohol consumption benefit economically from the trade and from tax revenues, but this may be outweighed by the costs of alcohol-related harm.

Learning outcomes for Chapter 1

After studying this chapter and its associated activities, you should be able to:

LO 1.1 Define and use in context, or recognise definitions and applications of, each of the terms printed in **bold** in the text. (Questions 1.1 to 1.7)

LO 1.2 Describe the main sources of uncertainty in estimates of the amount of alcohol consumed by individuals and populations. (Question 1.1 and DVD Activity 1.3)

LO 1.3 Summarise and illustrate the main trends in the amount of alcohol consumed, the patterns of drinking behaviour, and the distribution of alcohol-related harm in different regions of the world, with regard to gender and age-group. (Questions 1.4 and 1.5 and DVD Activity 1.3)

LO 1.4 Calculate a percentage and use metric measures of volume (l, cl and ml) correctly in estimating the amount of pure alcohol in an alcoholic drink. (Questions 1.2 and 1.3)

LO 1.5 Observe, summarise and note the main patterns in data presented in simple graphs and tables. (Questions 1.4 and 1.5)

LO 1.6 Describe the major types of acute and chronic effects on health that can result from the excessive consumption of alcohol. (Question 1.6)

LO 1.7 Explain why age-standardisation may be necessary before comparing health data from countries with different age structures. (Question 1.7)

LO 1.8 Discuss the impact of alcohol on individuals, families and on national economies. (Question 1.8 and DVD Activity 1.3)

If you are studying this book as part of an Open University course, you should also be able to:

LO 1.9 Judge when to 'round' decimal numbers to express them to one, two or three decimal places and to the nearest whole number. (Activity C1 in the *Companion*)

LO 1.10 Plot data from a simple table into a rough sketch graph on plain paper and use it to help you interpret the data. (Activity C2 in the *Companion*)

LO 1.11 State some of the pros and cons of using a search engine, combine keywords and phrases to search more effectively, and evaluate the quality of information on a website by using the PROMPT criteria. (Activity C3 in the *Companion*)

Figure 1.14 Label on a can of UK lager. (Photo: Lesley Smart)

Self-assessment questions for Chapter 1

You also had the opportunity to demonstrate LOs 1.3 and 1.8 by answering questions in DVD Activity 1.3.

Question 1.1 (LOs 1.1 and 1.2)

Why is it important to know the level of abstinence in a population and the proportion who are below a certain age when estimating the average consumption of alcohol per person per year?

Question 1.2 (LOs 1.1 and 1.4)

The label on a bottle of whisky says '40% vol.' A small glass of whisky contains 25 ml. What volume of pure alcohol does the glass contain in ml? How much alcohol is this in centilitres?

Question 1.3 (LOs 1.1 and 1.4)

Use the information shown on the label in Figure 1.14 to determine if the claim that it contains 2.0 UK units of alcohol is correct.

Question 1.4 (LOs 1.1, 1.3 and 1.5)

Figure 1.3 shows that in most of Western Europe annual alcohol consumption is in the *highest* band in the world (greater than 13 litres of pure alcohol per person aged 15+ years), whereas Figure 1.4 shows this region to be in the *lowest* risk category from alcohol drinking. Suggest an explanation for this observation.

Question 1.5 (LOs 1.1, 1.3 and 1.5)

Look again at Figure 1.6.

(a) What do you notice about the geographical distribution of binge drinking among young people in the countries shown?

(b) Compare the responses of boys and girls and sum up the main difference between them.

(c) Why are the data in Figure 1.6 a cause for concern?

Question 1.6 (LOs 1.1 and 1.6)

Explain why alcohol is ranked higher as a cause of disability-adjusted life years (DALYs) among *young* people in developed countries than it is among adults as a whole.

Question 1.7 (LOs 1.1 and 1.7)

If you wanted to compare the male death rate from alcoholic poisoning in the USA with the rate in Argentina, why it would be wise to age-standardise the data first?

Question 1.8 (LO 1.8)

Summarise the sources of financial loss to national economies resulting from excessive alcohol consumption.

CHAPTER 2

THE CHEMISTRY OF ALCOHOL

In the previous chapter, you examined the impact drinking alcohol has on global society in terms of health, cost and social breakdown. In the forthcoming chapters we take a closer look at what happens to alcohol in the body, and why it has such a strong effect on the body. To do this, we first need to introduce you to the basic chemistry surrounding these interactions.

2.1 Introduction

The 'alcohol' that is referred to in drinks could be termed in everyday language to be a particular 'chemical'. It is one of a family of similar chemicals that have the generic name of alcohol, and the particular one that is present in alcoholic drinks has the chemical name **ethanol**. Alcohol for drinking is made by the fermentation of carbohydrates such as natural sugars in grapes (wine), or starch in grains (beer, saki) using yeast. It can be concentrated and purified by distillation to produce spirits such as brandy, whisky, and vodka.

When alcohol enters the body, as with any food, the body reacts to break it down and make it useful. Thus alcohol undergoes chemical reactions in the body, which change it in stages, eventually forming carbon dioxide and water and releasing energy. To understand these reactions you need to know more about atoms, elements and molecules.

Fermentation is the process whereby a fungus (yeast) acts on sugars and converts them into ethanol and carbon dioxide gas which bubbles off. *Distillation* is a method of separating liquids with different boiling temperatures, here water and ethanol.

2.1.1 Atoms and elements

An **element** is a simple substance which cannot be broken down into simpler substances by chemical reaction. There are only 92 naturally occurring elements found on the Earth; they each consist only of the smallest particles that are *characteristic of that element*, **atoms**. For instance, the metal sodium is an element and consists only of sodium atoms; hydrogen, a gaseous element, contains only hydrogen atoms.

Atoms are made up of several different sorts of smaller particles. The three most important here are **protons, neutrons** and **electrons**. The **atomic nucleus** contains protons, each of which carries a unit of *positive* electric charge, and neutrons which are uncharged. The atoms of different elements have different numbers of protons. The atomic nucleus is surrounded by electrons; electrons are about 2000 times lighter than protons and neutrons, but each one possesses a unit of *negative* charge. An atom has no overall electric charge because the number of protons in an atom equals the number of electrons, and so the positive charge of the nucleus is exactly balanced by the surrounding electrons each of which carries a negative charge equal and opposite to the charge of a proton.

Early models (Figure 2.1 proposed by Ernest Rutherford) of the atom likened the motion of the electrons around the atomic nucleus to that of the planets in orbits around the sun. It is now known that the motion of electrons is more complex than this, but for our purposes it is sufficient to understand that the electrons are in motion at a distance from the nucleus.

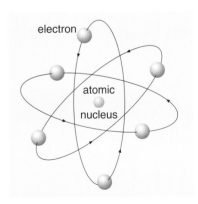

Figure 2.1 The Rutherford model of the atom, showing a central nucleus with six electrons moving around it.

Each element has been assigned a **chemical symbol**, many of which are easy to remember as they are the first letter, or two of the first letters of the name. For instance H stands for hydrogen, C for carbon, N for nitrogen, Ca for calcium and O for oxygen.

2.1.2 Bonding between atoms

Atoms can join together to form more complex structures or **compounds** such as water, H_2O, a compound of hydrogen and oxygen. Atoms are said to form chemical **bonds** with each other, or that they are 'bonded', and this is due to electrical forces between the atoms. Two main types of bonding are called *ionic* and *covalent* bonding.

Ionic bonding and ionic compounds

A metal atom can lose an electron (sometimes more than one) which is itself donated to the atom of another element.

◈   What is the charge on a metal atom if it loses an electron?

◆   The metal atom now has a *positive* charge.

The atom receiving the electron becomes *negatively* charged. This process of electron loss or donation is referred to as **ionisation** and the charged particles formed are known as **ions** (eye-ons). In sodium chloride, chemical formula, NaCl, the compound known as common salt (the white crystals you use on food and in cooking), the sodium atom easily loses one electron to form the positively charged sodium ion, Na^+. The positive charge is shown as a *superscript on the right* of the chemical symbol. The electron lost from sodium is transferred to a chlorine atom, which becomes negatively charged, forming the chloride ion, Cl^-, with the negative electrical charge also shown as a superscript.

The ions now have equal and opposite electrical charges and so attract one another and it is this electrical attraction which holds the salt crystals together. The bonding in this case is referred to as **ionic bonding**. Ionic compounds are frequently called by the generic term '*salts*'.

Covalent bonding and molecules

In other compounds, instead of an electron being transferred completely from one atom to another, an outer electron from each of two adjacent atoms is shared between the two atoms, forming an electron-pair bond. This type of bonding is called **covalent bonding** and is used in forming **molecules**. Molecules vary tremendously in size from the smallest containing only two atoms, to molecules such as proteins which contain thousands of atoms. The simplest example of a molecule is where only two identical atoms are bonded together. Take the example of hydrogen gas. Hydrogen is an element and so contains only hydrogen atoms, which each possess only one electron. Under normal conditions hydrogen

The symbols for a few elements are less obvious as they derive from an older name or different language, such as Na for sodium (Latin, *natrium*), K for potassium (north African Arabic, *kali*), W for tungsten (German, *wolfram*) and Cr for chromium (Greek, $\chi\rho\omega\mu\alpha$ (*chroma*)).

Positively charged ions are known as *cations* and negatively charged ions as *anions*.

Common salt (NaCl) is the most familiar example of a salt, but there are many others.

does not exist as single atoms, but two hydrogen atoms bond together, *sharing* their electrons in an electron-pair bond and forming a molecule, H_2,

$$H \overset{\text{x}}{\underset{\text{x}}{}} H \quad \text{— electron pair}$$

where × represents the hydrogen electrons – one from each atom.

The electron-pair bond is often referred to as a covalent bond. In hydrogen it is often depicted as H—H, where the straight line denotes the covalent bond between the two atoms.

In summary:

(i) Atoms bond together to form chemical compounds.

(ii) There are two important types of chemical compounds which can be distinguished by their bonding:

- **ionic compounds** (often called salts), which contain positively and negatively charged ions and

- **molecular compounds** which contain covalently bonded molecules.

2.1.3 Formulae

There is a shorthand way of denoting the composition of a chemical compound – the **chemical formula** (often called the 'formula'). The formula tells you:

- which type of atoms are bonded together to make up the compound, and

- how many atoms of each type there are – this is given as a *subscript* to the right of the chemical symbol for the element.

The word formulae (form-you-lee) is the Latin plural of formula. Although common usage can add an 's' to make the plural – formulas – this has not been generally adopted by scientists and mathematicians.

Taking some examples of molecular compounds, for hydrogen gas the formula H_2 indicates that there are only hydrogen atoms in the molecule and that there are two of them. The gases oxygen and nitrogen have the formulae O_2 and N_2 respectively, meaning that they each contain two atoms of the element bonded together to make a molecule. Other familiar molecules are water and carbon dioxide, with the formulae H_2O and CO_2 respectively. H_2O indicates that the molecule contains two atoms of hydrogen and one of oxygen bonded together. CO_2 contains one atom of carbon and two atoms of oxygen.

$$H_2 \quad O_2 \quad N_2 \quad H_2O \quad CO_2 \quad H_2O$$

subscripts

The formula does not show in what *order* the atoms are joined together; for this information chemists turn to a **structural formula**. You could imagine two possibilities for the water molecule:

$$H—H—O \quad \text{or} \quad H—O—H$$

terminal oxygen   central oxygen

Experiments have shown that the second structural formula, with the oxygen atom between the hydrogens, represents the true bonding situation.

2.1.4 How many bonds can an atom make?

Not all the electrons in an atom are used for bonding to other atoms, as the inner electrons are tightly bound to the atomic nucleus and so are unavailable to form bonds. Bonds are formed with other atoms using the outermost electrons only. The number of outer or *bonding* electrons available varies from element to element.

Ethanol is a covalently bonded molecule which contains only the elements hydrogen, oxygen and carbon. These (together with nitrogen) are the most important elements involved in the chemistry of the body, and each one has a different number of bonding electrons.

In hydrogen, H_2, each hydrogen atom has a single electron which is shared with the other hydrogen atom to form one covalent, electron-pair, bond between the two atoms. This is called a **single bond** and depicted as H—H.

In water, H_2O, each hydrogen atom shares its electron with one electron from oxygen to make a single bond to oxygen, H—O—H. The oxygen atom is able to share *two* electrons, one to each hydrogen, forming *two* single bonds.

In oxygen, O_2, each oxygen atom is able to share *two* electrons with the other oxygen. This sharing of two electrons from *each* atom forms a **double bond**, depicted as O=O.

◆ In carbon dioxide, drawn as O=C=O, carbon forms a double bond with each oxygen atom. How many electrons does the carbon atom use in bonding?

◆ Carbon has four bonding electrons which pair-up in twos with two bonding electrons on each oxygen atom.

2.1.5 Shapes of molecules

One piece of information that is not always easy to show on the flat printed page is the *shape* of the molecule. A computer model provides a very useful way of looking at the 3D nature of molecules, and the molecules O_2, CO_2 and H_2O are modelled in Activity 2.1.

Activity 2.1   The shapes of simple molecules

Allow 30 minutes

Now would be the ideal time to study the DVD activity entitled 'The shapes of simple molecules' on the DVD associated with this book. If you are unable to study it now, continue with the rest of the chapter and return to it as soon as you can.

Interactive computer models of the molecules O_2, CO_2, and H_2O can be found on the DVD. The exercises allow you to examine their shape and bonding.

2.2 The alcohol molecule

The proper chemical name for the alcohol in alcoholic drinks is ethanol. The chemical formula of ethanol is usually written as C_2H_5OH (spoken as sea-two-aitch-five-oh-aitch).

◆ How many atoms of each type are there in ethanol?

◆ There are two carbon atoms, six hydrogen atoms, and one oxygen atom.

◆ How many atoms in total does this molecule contain?

◆ There are *nine* atoms in total.

◆ Does this formula tell you the order in which the atoms are bonded?

◆ The formula does *not* tell you the order in which the atoms are bonded. For that you need the *structural formula*, which is described in the next section.

2.2.1 The structural formula of ethanol

The way in which atoms bond together determines the size and shape of a molecule, and how the electric charge is distributed over its surface. These factors in turn determine the properties of a molecule, for instance whether or not it can dissolve in water, whether it is small enough to move through a membrane, or whether it is able to fit into a receptor site, like a key into a lock. Examining the properties of ethanol allows a more detailed look at its interactions throughout the body in later chapters.

The structural formula of ethanol shows the order in which the atoms are joined together. Note that there are only *single* bonds.

◆ Look at each atom in turn. Is it bonded to the expected number of other atoms? (Look back at Section 2.1.4 if you can't remember the number of bonding electrons for each element.)

◆ Yes. The oxygen atom makes two single bonds – one to carbon and one to hydrogen. Each carbon atom is surrounded by four single bonds, and each hydrogen atom has one.

As you saw in Activity 2.1, the structural formula does *not* show you the shape of the molecule.

The older name for ethanol is ethyl alcohol and this is still frequently used.

The formula for ethanol is usually written C_2H_5OH, rather than C_2H_6O as it distinguishes differently bonded hydrogen atoms.

You will now find both alcohol and ethanol used in this book. Where possible, we have tried to use 'alcohol' to refer to an alcoholic drink, and 'ethanol' to refer to the chemical and its effects.

2.2.2 The shape, size and structure of ethanol

The size, the shape (Figure 2.2) and the electric charge distribution on ethanol dictate where and how it can move in the body. The size of a molecule affects whether it can move through a cell membrane and the speed at which it can do it. Ethanol for instance moves quickly through the walls of the stomach and small intestine into the bloodstream; a process taken up in more detail in Chapters 3 and 5.

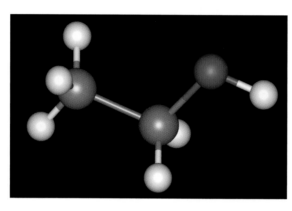

Figure 2.2 Computer model of the ethanol molecule. (White, H; grey, C; red, O) (Accelrys)

The ability of ethanol to dissolve completely in water to form a homogeneous liquid or *solution*, enables it to be carried around easily in the bloodstream and thus taken to any organ (Chapter 3). But *why* is ethanol able to dissolve completely in water? Many chemicals, oil and petrol (gasoline) for instance, can't. Understanding more about the distribution of the bonding electrons and thus of the distribution of charges on the molecule, explains why this happens.

In order for a signalling molecule to fit into a *receptor*, like a key into a lock, not only is the size and shape of the molecule very important, as it has to fit the shape of the receptor site, but also the distribution of the bonding electrons is important. Small changes in the position of the charges on the surface of a molecule can attract the molecule into the site and hold it with weak bonding. As you will see in Chapter 4, the ethanol molecule is able to attach itself to receptors on certain cells in the brain and trigger the release of signalling molecules called neurotransmitters (see Section 4.3.1).

The following multimedia exercise explores the structure of ethanol.

Activity 2.2   The structure of ethanol

Allow between 30 minutes and 1 hour

Now would be the ideal time to study the DVD activity entitled 'The structure of ethanol' on the DVD associated with this book. If you are unable to study it now, continue with the rest of the chapter and return to it as soon as you can.

This activity allows you to explore the structure and shape of the ethanol molecule and the geometry around each atom. You will use a tool to measure the distances between atoms. It also depicts the electron surface of the molecule, reflecting the charges on the atoms, and shows how this leads to its solubility in water, comparing it with how salts such as NaCl dissolve in water.

A further example of the importance of the shape and bonding in molecules is described in Box 2.1.

Box 2.1 (Enrichment) Modern drug research

Much modern pharmaceutical research is based on computer molecular modelling techniques to design new drugs, modelling their size, shape and electric charges in the way that you have seen displayed for the ethanol molecule in Activity 2.2. For example, great success was achieved in this way with the anti-flu drug Relenza® (Figure 2.3). The enzyme sialidase is found in the outer coating of the flu virus and is involved in the spread of the virus. Relenza inhibits the operation of sialidase.

Enzymes and their role are discussed in more detail in Section 2.4.2.

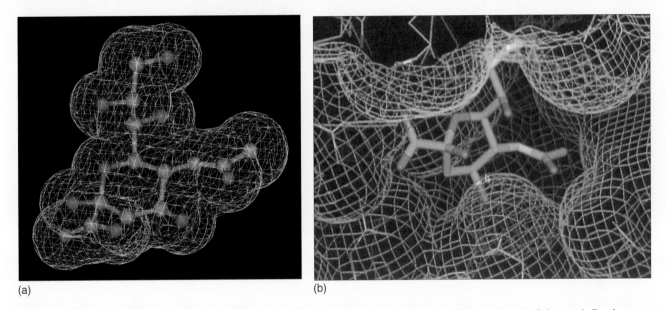

(a)   (b)

Figure 2.3   (a) Computer model of sialic acid, a molecule important in the development of the anti-flu drug Relenza®, shown with the electron distribution surface, and (b) the receptor site for sialic acid in the enzyme sialidase. (Source: Guy Grant)

2.3   The properties of ethanol

2.3.1 Solubility in water and the importance of hydrogen bonding

To recap what has been covered so far, you saw in Activity 2.2 that ethanol is similar to water in that they both have the hydroxyl or −OH (oh-aitch), group of atoms in the molecule.

Note that in the hydroxyl group, chemists often don't bother to draw in the bond between the O atom and the H atom (—OH), but *do* put in the bond which is connected to the rest of the molecule.

The Greek lower-case delta, δ, is conventionally used throughout maths and science to represent a small or partial quantity.

You are now familiar with the idea that a chemical bond is formed either when atoms share their outer electrons with each other, or, in the extreme, when they donate an electron to another atom thus forming an electrical attraction between the ions so formed. Some atoms are better at attracting electrons to themselves than others: oxygen is one of these, and it is said to be **electronegative**. In a hydroxyl group the oxygen atom tends to pull rather more than its fair share of the electrons in the bond towards itself, making it slightly more negatively charged, whereas the H atom becomes slightly positively charged in response.

When writing out formulae on paper, chemists *depict* the drift of negatively charged electrons towards a more electronegative element such as oxygen, by writing over the oxygen a *partial* negative charge, δ− (delta-minus), and over the hydrogen atom it is attached to, a partial positive charge, δ+ (delta-plus):

oxygen has
a slight negative
charge   hydrogen has
       a slight positive
$$\overset{\delta-}{}\quad\overset{\delta+}{}$$
$$C_2H_5 - O - H$$
charge

where the δ− indicates that the bonding electrons spend more time near the oxygen atom and less time near the hydrogen atom (δ+). (Note that this does *not* mean that the electron is divided in any way).

One of the effects of this separation of charge within a molecule is that neighbouring molecules can form weak bonds with each other due to the attraction of the opposite charges. This happens particularly strongly in water itself, where the $H^{\delta+}$ of one molecule is attracted to the $O^{\delta-}$ of a neighbouring molecule. This is known as **hydrogen bonding**, and is illustrated in Figure 2.4.

The separation of charges on the —OH groups of both ethanol and water molecules means that these two molecules also form weak hydrogen bonds with each other, and so ethanol dissolves in water (Figure 2.5).

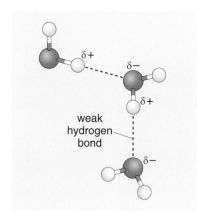

Figure 2.4 Hydrogen bonding between water molecules. (Red, O; white, H)

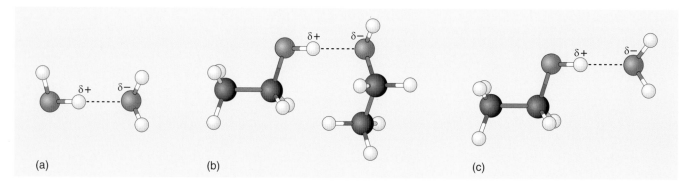

Figure 2.5 Ethanol and water molecules showing the weak attraction to each other due to the partial charges on their —OH groups. (a) Hydrogen bond between two water molecules; (b) hydrogen bond between two ethanol molecules; (c) hydrogen bond between one ethanol and one water molecule. (Red, O; white, H; grey, C)

Because of the solubility of ethanol in water, when a person drinks an alcoholic drink, the ethanol also dissolves easily in the blood and body fluids, and so gets transported all around the body, affecting each organ as it goes.

It is difficult to overstate the importance of hydrogen bonding to biological systems, as without it no forms of life would exist. Hydrogen bonding has a dramatic effect on the properties of water – without it, water would be a gas at normal temperatures, and the oceans, rivers and lakes would not exist; it is also responsible for the structure of ice, as described in Box 2.2 overleaf.

Hydrogen bonding holds the two strands of DNA together (Figure 2.6). The two helices of the backbone of the DNA molecule are shown, one in orange and one in blue; the helices are joined to each other by dotted lines which represent the hydrogen bonds between the two strands.

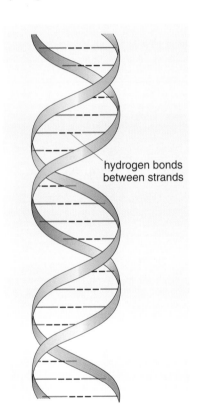

hydrogen bonds
between strands

Figure 2.6   DNA is the huge double-helix molecule found in the nuclei of almost all cells in the body (but not red blood cells). It carries the genetic code containing the instructions to construct other components of the cell such as proteins. The two strands of the double helix are held together by hydrogen bonding.

Box 2.2 (Enrichment) Why do fish survive in a frozen lake?

You will probably have noticed that ice floats on water (Figure 2.7a, b); this phenomenon is also a direct consequence of hydrogen bonding between water molecules (Figure 2.7c).

In general when a liquid cools and becomes a solid, it becomes more dense (that is, the same volume weighs more). If a liquid forms some crystals, you will notice that they usually sink to the bottom of the container; you sometimes notice this happening in olive oil, or runny honey. A corollary of this is that as you cool a liquid down, it usually contracts – as it becomes denser, it takes up less volume. Water is very unusual because it does the *opposite* of this at low temperatures. Water turns into ice at 0 °C, but it is most dense at 4 °C. As the water cools below

4 °C, the hydrogen bonds make a continuous rather open, three-dimensional cage-like network which as a consequence takes up a *larger* volume; so as the water crystallises into ice, the ice is *less* dense than the water. You may have had bitter experience of this if you have put a bottle of wine or beer to cool in the freezer and forgotten it – it is quite possible for the bottle to burst due to the expansion of the water. Similarly, cartons of milk stored in the freezer become very 'bloated'.

Fortunately for the fish, as ice forms on a lake, it floats to the top because it is less dense, providing an insulating layer, and the fish can still swim around happily underneath. Were it the other way round, they would be left on top of the ice.

(a) (b) (c)

Figure 2.7 (a) Iced water. (Photo: Lesley Smart). (b) Frozen lake. (Photo: Scott Robinson/Flickr Photo Sharing). (c) Structure of ice: the short H—O distances are covalent bonds, the long dashed ones are hydrogen bonds.

2.3.2 Boiling temperature and vapour pressure of ethanol

Ethanol has different properties from water because attached to the —OH group, instead of a single hydrogen atom, H—OH, there is an ethyl group, C_2H_5—OH.

For instance pure ethanol freezes at a lower temperature than water: water freezes at 0°C, whereas pure ethanol freezes at −117.3 °C. Even 40% vol. spirits such as gin and tequila, popularly drunk very cold, freeze at a sufficiently low temperature that they can be kept in the freezer at −18 °C and remain liquid. This is not true for drinks with a lower ethanol content such as wine (12–15%) or beer (3–5%), which will freeze and burst their containers.

Ethanol also boils at a lower temperature than pure water, 78.5 °C compared with 100 °C. At its boiling temperature a liquid changes state to become a gas. Because molecules are constantly in motion, at any moment the faster molecules are leaving the surface of the liquid to become a gas (evaporation) and others are returning (condensation) to the liquid. Thus even at temperatures *below* the boiling temperature, there are always a certain number of gaseous molecules in the air above a liquid. These gaseous molecules now become partially responsible for the atmospheric pressure on Earth and are said to form a *vapour pressure.* As the temperature of the liquid is raised, the molecules move faster and more molecules vaporise into the surrounding air. The vapour pressure of a liquid thus increases with temperature. Normal body temperature is 37 °C, and at this temperature ethanol has a sufficient vapour pressure in the breath expelled from the lungs to allow detection using a breathalyser. This is also why you can smell 'alcohol' on the breath of someone who has been drinking. Box 2.3 explains how breathalysers work and you will see one in use later in Activity 4.2.

> The gas above a liquid is commonly called a vapour.

Box 2.3 (Explanation) Breathalysers

The ethanol in alcoholic drinks is absorbed into the bloodstream through the gut, and is carried around the body in the bloodstream to the organs. As it moves in the blood passing through the lungs, some of it diffuses into the air in the lungs and is breathed out; it is this ethanol that is detected by breath-alcohol detecting devices, commonly known as breathalysers. (Diffusion is explained in Chapter 3.)

A person suspected of drinking and driving is asked to blow into the detector device (Figure 2.8 overleaf), and the concentration of ethanol in this sample of exhaled air is measured and gives an estimate of the corresponding ethanol level in the blood. Most countries have a drink-driving limit for the blood-ethanol concentration varying from 0 to 90 mg per 100 ml of blood; a comparative list is given in Table 5.1. In the UK, the limit is one of the highest at 80 mg/100 ml of blood. At 37 °C this is equivalent to 35 µg/100 ml of exhaled air (the concentration will vary slightly with body temperature). If a person in the UK gives a breathalyser reading *above* 35 µg/100 ml exhaled air they are offered a blood test to see if their blood-ethanol concentration is above 80 mg/100 ml blood. If they refuse the blood test, they can be charged with drink-driving on the basis of the breathalyser result alone.

> mg per 100 ml is often abbreviated to mg/100 ml.

> If you are unfamiliar with the units, the table in Appendix 1 lists some of the prefixes used for SI units with some common examples.

Figure 2.8 A breathalyser device in use. (Photo: Jack Sullivan/Alamy)

The blood-ethanol concentration is known as **BAC**, **blood-alcohol concentration**, and is usually quoted in mg/100 ml of blood. Note that the concentration is quoted as the *mass* of ethanol (in milligrams) per volume of blood or air, whereas the drinks industry quotes alcohol concentrations as a percentage *volume* (in cl or ml). There is no particular advantage to either method.

One way in which these devices operate is by using a chemical that undergoes a colour change. Potassium dichromate undergoes a reaction with ethanol to form acetic acid (ass-ee-tik); in so doing, it changes from orange to green, a change that can be measured quantitatively. Most modern breathalysers give a digital display of the BAC.

2.4 The metabolism of ethanol in the body

The ability of ethanol to dissolve in water allows it to make its way from the gut into the bloodstream, and from there it is transported throughout the body. Very little of the ethanol taken in is excreted unchanged; about 10% is expelled without reaction, of which approximately 5% passes through the kidneys and is excreted in the urine; a similar amount is breathed out from the lungs, and a little is lost in sweat. The remaining 90% of the ethanol is converted into other chemicals; this process of breaking down an ingested food into other molecules, releasing the energy needed to maintain life, is part of the body's **metabolism**. For ethanol this takes place in the liver.

2.4.1 The breakdown of ethanol in the liver

Ethanol is metabolised (broken down) in the liver in three stages. First acetaldehyde (ass-et-al-dee-hide), CH_3CHO, is formed from ethanol. This process can be abbreviated using formulae and an arrow to symbolise that a chemical transformation has taken place:

$$C_2H_5OH \longrightarrow \underset{\text{acetaldehyde}}{CH_3CHO} \tag{2.1}$$

The internationally accepted chemical names for acetaldehyde and acetic acid are *ethanal* and *ethanoic acid*, respectively. They are not always used and here the traditional, more familiar names are used. The reactions have been numbered (2.1), (2.2), etc. so we can refer back to them easily.

The acetaldehyde then produces acetic acid, CH_3COOH:

$$CH_3CHO \longrightarrow \underset{\text{acetic acid}}{CH_3COOH} \tag{2.2}$$

and finally the acetic acid is broken down into carbon dioxide, CO_2 and water, H_2O:

$$CH_3COOH \longrightarrow CO_2 + H_2O \tag{2.3}$$

Note that these representations are not chemical *equations* as they do not 'balance' – there are different numbers of atoms on either side of the arrows. We cover this in Section 2.5.1.

Overall ethanol breaks down to carbon dioxide and water. This is also the reaction that takes place when ethanol burns in the air, but in burning it produces carbon dioxide and water vapour in a *single* fast reaction and does not go through the two intermediate stages of forming acetaldehyde and acetic acid. You can see an example of this reaction in the home if you set fire to spirits poured over a pudding or in a drink (Figure 2.9).

To burn ethanol in the air, a large 'kick start' is needed by way of a lighted match or a spark; once it has started to burn, a very lively chemical reaction takes place producing a great deal of heat and light (Box 2.4). We need to be very thankful that this is not the way that the reaction takes place in the body!

Figure 2.9   Brandy burning on a Christmas pudding – an example of the combustion of ethanol. (Photo: Matt Rigott/Flickr Photo Sharing)

Box 2.4 (Enrichment) A sad tale

In 14th century Europe, brandy was thought of as somewhat of a miracle cure for all ills. Charles the Bad (Charles II of Navarre) thus came to a terrible end in 1387: his doctors sewed him into a brandy-soaked sheet, and when a servant held a candle too close, the sheet and the patient went up in flames.

The conditions in the cells of the liver are very different from those in the open air, and under these warm, enclosed conditions with ethanol in solution in water, it would not react at all were it not for the presence of some large protein molecules known as **enzymes**. An enzyme is a molecule that has the ability to accelerate a *particular* chemical reaction in a cell, allowing it to take place at body temperatures. Enzymes remain unchanged at the end of the reaction. Molecules such as this, which make a reaction go faster but which can be recovered unchanged at the end of the reaction, are known as **catalysts**; in the cell they bind to their molecule of action (substrate), act upon it to facilitate the reaction for which they are responsible, and then leave the now changed substrate – the product of the reaction – and move on to the next substrate.

In the cells, ethanol is first converted into acetaldehyde through a chemical reaction facilitated by the enzyme, *alcohol dehydrogenase,* ADH. Acetaldehyde is a more toxic chemical than ethanol, and if it builds up too much in the bloodstream because of excess drinking, it causes people to feel very unwell.

ADH can also be used as the abbreviation for anti-diuretic hormone. We will not use it for this in this book, to avoid any confusion.

ADH is not a single enzyme: it exists in five different forms called *isoforms*. Which ADH isoforms are present in a particular individual depends on the genetic makeup of that person. Two of the five types of ADH metabolise ethanol to acetaldehyde more rapidly, resulting in the accumulation of higher amounts of acetaldehyde and making a drinker who possesses these forms feel uncomfortable more quickly – even a small amount of alcohol makes them feel very ill. This is common in people of Asian origin, but does have the positive effect of helping to protect them from developing alcoholism, a disease rare in Asia.

A problem for other individuals is that an *insufficient* amount of ADH is produced in the liver, and so that person will only be able to metabolise ethanol very slowly, causing the ethanol to remain in the system longer, thus prolonging intoxication. This tends to happen more as people get older.

In the final two steps of the ethanol breakdown, acetaldehyde is quickly converted by another liver enzyme, called *aldehyde dehydrogenase,* ALDH, into acetic acid, a non-toxic molecule in humans. As you will see in Section 5.1, ALDH, like ADH, also exists in several isoforms.

Acetic acid is finally broken down into water and gaseous carbon dioxide, CO_2, which is eliminated through the lungs. The drug sometimes given to help recovering alcoholics, Antabuse®, works on the basis that it prevents this final reaction from taking place, thus keeping the levels of toxic acetaldehyde high. If a person drinks alcohol after taking Antabuse, it makes them build up too much acetaldehyde in the system and thus feel very sick.

Elimination of CO_2 through the lungs is discussed in another book in this series *Chronic Obstructive Pulmonary Disease: A Forgotten Killer* (Midgley, 2008)

The liver can break down only a certain amount of ethanol per hour, regardless of the amount that has been consumed. If the rate of alcoholic drink consumption exceeds the rate at which the ethanol can be metabolised, then the concentration of ethanol in the blood increases and the individual may become intoxicated. Figure 2.10 shows a typical graph for the change in blood-alcohol concentration, after one standard USA drink (see Table 1.4).

◈ How long does it take for the ethanol concentration to maximise?

◆ BAC peaks around 50 minutes.

BAC generally peaks at some time between 45 and 90 minutes after consumption, with the maximum dependent on the amount consumed. The rate of ethanol metabolism depends on many factors which vary from person to person: the isoforms and amounts of ADH and ALDH in the liver, body mass, gender, amount and type of beverage ingested, duration of drinking, fatigue, and the presence and type of food. Some of these factors are explored in Chapter 5.

◈ In Figure 2.10, look at the slope of the curve on either side of the maximum. What does it tell you about the relative *rates* of absorbing ethanol into the body and metabolising it?

◆ Ethanol is absorbed quickly into the body as shown by the steep slope on the left. The shallower slope to the right during elimination of ethanol, shows that it is metabolised much more slowly than it is absorbed into the body.

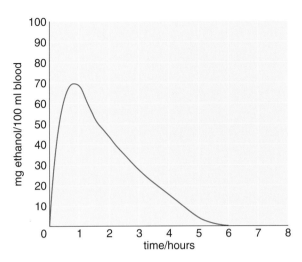

Figure 2.10 A typical graph of the change in blood-alcohol concentration (BAC) with time after consumption of a standard USA drink (17.7 ml ethanol). (Source: National Institutes of Health, 2003)

The rate of metabolism ranges in different individuals from a decline in BAC of less than 10 mg of ethanol per 100 ml of blood per hour to over 40 mg per 100 ml per hour (i.e. about 1 UK unit in 3 hours to 1 UK unit in less than an hour).

Since the metabolism of ethanol is slow, consumption needs to be controlled to prevent accumulation in the body and the resulting intoxication.

These complex processes of metabolism, with alcohol as with all foods, are what provide the energy for the body to keep working. In order to understand how this energy is generated you now need to understand chemical reactions in a little more depth, and it is to this that we turn in the next section.

2.5 The chemical reactions of ethanol

When something happens to a chemical compound and a change of some description takes place, there has been a *chemical reaction*. Before looking at the reactions of ethanol, you will first see how chemists express a chemical reaction in the form of a *chemical equation*.

2.5.1 Balancing chemical equations

You learnt in the last section that the metabolism of ethanol is equivalent overall to burning it in the oxygen of the air, so in this section on equations we'll take some similar combustion reactions with oxygen as examples, starting with the reaction to make water. This uses two elements that you came across earlier: hydrogen gas, H_2, and oxygen gas, O_2, which react with each other (explosively) to form water, H_2O, and nothing else. The simplest way of writing this down chemically is to write down the formulae of the elements that react with each other, called the *reactants*, and, as before, use an arrow to indicate what has been formed in the reaction – the *products*, thus:

$$\overbrace{H_2 + O_2}^{\text{reactants}} \longrightarrow \overbrace{H_2O}^{\text{product}}$$

(2.4)

Notice the convention of reactants on the left and products on the right. This equation is useful in that it tells you, in a shorthand form, which type of molecules you start with and finish with, and you can think of it in terms of the molecules interacting. However, if you look closely at the equation, you'll see that there is a problem in that the number of atoms in the reactants is not equal to the number of atoms in the products, and *atoms cannot be destroyed in a chemical reaction*. Chemists would say that this equation is 'not balanced'. It is important to be able to produce a balanced equation for a chemical reaction, because it is only from this that the *quantities* of the substances involved can be calculated, and it is from these that the energy absorbed or released by a reaction can be determined – the ultimate goal of this chapter is to calculate this for the combustion of ethanol.

Starting with the first element, hydrogen: the hydrogen atoms balance – there are two on either side of the arrow. Moving to oxygen, there are two oxygen atoms on the left, but at present only one on the right. As an atom of oxygen cannot be lost, and only water has been produced, *two* molecules of water must have been made. To indicate this, a *prefix* '2' is placed in front of the formula for the water molecule.

Notice that when the number of a particular molecule taking part in a reaction is one, the prefix '1' is omitted.

$$\text{prefix} \searrow$$
$$H_2 + O_2 \longrightarrow 2H_2O \tag{2.5}$$

Unfortunately this immediately unbalances the equation again as now there are two hydrogen atoms on the left, and four (2×2) on the right, and this is where you must hold your nerve – you simply go through the process again! This time, taking two reactant molecules of hydrogen corrects the problem:

$$2H_2 + O_2 \longrightarrow 2H_2O \tag{2.6}$$

And now the equation 'balances'. When the equation balances the arrow can be replaced with an equals sign. A final refinement, not always used, is to put the physical state of the reactants and the products in brackets after their formulae – **g** for gas, **l** for liquid, **s** for solid, and **aq** for a substance dissolved in water (aqueous solution). The finished equation is thus:

$$2H_2(g) + O_2(g) = 2H_2O(l) \tag{2.7}$$

prefix / subscript / denotes a liquid–in this example water is in the liquid form

Notice the different uses of numbers in the equation. The *subscript 2* in the formula indicates that two hydrogen atoms are contained in the water molecule. The *prefix 2* in front of the water formula indicates that there are two H_2O molecules.

Another example of a simple reaction involving oxygen, is the burning of so-called 'natural gas', a chemical reaction you see if you use a gas hob in a kitchen. The chemical reaction involved is the reaction of the oxygen gas, O_2, from the air, with methane, CH_4, to produce carbon dioxide and water vapour. (Note that because this reaction takes place at high temperatures, above the boiling temperature of water, the water produced is a gas (water vapour) not a liquid.)

◈ Write down the reactants and products using words and then chemical formulae, but don't try to balance the equation yet.

◆ The equations are:

$$methane + oxygen \longrightarrow carbon\ dioxide + water \qquad (2.8a)$$

$$CH_4 + O_2 \longrightarrow CO_2 + H_2O \qquad (2.8b)$$

If there is more than one product, as in this case, the procedure for balancing the equation is the same as before, it just usually requires a little more patience. Balance each element in turn, starting on the left with carbon. In Equation 2.8b, carbon is already balanced, there is one carbon atom on each side. Next consider hydrogen: there are four on the left and only two on the right, so there must be two molecules of water formed:

$$CH_4 + O_2 \longrightarrow CO_2 + 2H_2O \qquad (2.9)$$

Finally the oxygen atoms must be balanced. There are now two on the left and four in total on the right (Equation 2.9) so two molecules of oxygen on the left will do the trick. The final equation with the physical states, is:

$$CH_4(g) + 2O_2(g) = CO_2(g) + 2H_2O(g) \qquad (2.10)$$

The reactions shown in equations 2.7 and 2.10, involving a reaction with oxygen, are particular examples of **oxidation reactions**.

There are more examples for practising balancing equations in Questions 2.8 and 2.9 at the end of the chapter.

2.5.2   Burning ethanol

The final products of the metabolism of ethanol in the liver are carbon dioxide and water – the same products obtained from combustion, i.e. burning, ethanol in the oxygen in the air.

◈ Try to produce a balanced equation for the combustion of ethanol. Start by writing out the reactants and the products. Then balance each type of element in turn. All the reactants and products are gases, so we can leave out the physical states.

◆ The reactants and products are:

$$C_2H_5OH + O_2 \longrightarrow CO_2 + H_2O \qquad (2.11)$$

◆ The two carbons of ethanol must make two molecules of carbon dioxide:

$$C_2H_5OH + O_2 \longrightarrow 2CO_2 + H_2O \qquad (2.12)$$

◆ The six hydrogens of ethanol must produce three molecules of water:

$$C_2H_5OH + O_2 \longrightarrow 2CO_2 + 3H_2O \qquad (2.13)$$

◆ There are now seven oxygens on the right-hand side and so three molecules of oxygen are needed on the left to balance this:

$$C_2H_5OH + 3O_2 = 2CO_2 + 3H_2O \qquad (2.14)$$

Notice that if you chose to start with oxygen in Equation 2.11, then initially the oxygens balance; it is important that you still go on to balance each of the other elements until *all* are balanced.

During the combustion of ethanol (Figure 2.9) a great deal of heat is generated, or put another way, the reaction releases a lot of energy. Burning ethanol in air carries out in one fast high-temperature reaction, what the liver does in three low-temperature stages. The next section explores how much energy is generated in this reaction and compares the result with other foods and fuels. Another oxidation reaction of ethanol that produces a familiar 'food' is described in Box 2.5.

Box 2.5 (Enrichment) Vinegar from wine

Vinegars, which are simply dilute aqueous solutions of acetic acid, can be prepared from many different alcoholic beverages, but traditionally from wines and apple cider (Figure 2.11). If a bottle of wine is left uncorked it becomes vinegary; this is due to the formation of acetic acid from ethanol by the action of bacteria called *Acetobacter* enabling the following reaction to take place:

$$C_2H_5OH\,(aq) + O_2(g) = \underset{\text{acetic acid}}{CH_3COOH\,(aq)} + H_2O(aq) \qquad (2.15)$$

Acetobacter must be prevented from growing in wines and turning them to vinegar, and so wine must be bottled in very clean conditions, air must be excluded by good corking, and a preservative, sulfur dioxide, is usually added.

Figure 2.11   Different vinegars. (Photo: Lesley Smart)

2.6 Does drinking alcohol make people fat?

You may have heard that alcoholic drinks are very 'calorific' and therefore to be avoided when dieting. To answer the question in the title of this section, you first need to find out how much energy is released when ethanol is metabolised by the body, comparing it with the energy released, weight for weight, with the usual foods that people eat (Figure 2.12), and then consider briefly the complex question of how energy is stored in the body. The units of energy in common use are described in Box 2.6.

Box 2.6 (Explanation) Energy and its units

Energy exists in various forms which can be changed from one to another, such as heat, light and motion. It can also be stored, when it is known as *potential energy*. The internationally accepted (SI) unit for energy is the *joule*, symbol J. One thousand joules is 1 kilojoule, kJ. The older energy unit, still sometimes used by the food and dieting industry, is the *calorie*, symbol cal. One calorie is approximately equivalent to 4.2 joules. To confuse things further, the unit commonly quoted in diets, the 'Calorie' with an upper case 'C' is one thousand calories, or a kilocalorie, kcal. Notice the energy values on these food labels.

(a)

(b)

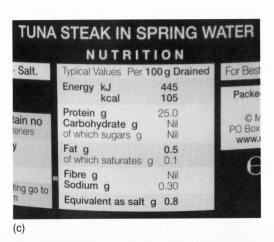

(c)

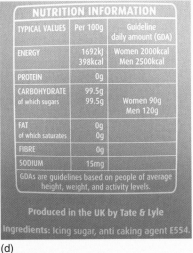

(d)

Figure 2.12 Energy ratings on some food labels. (Photo: Lesley Smart)

Chemical bonds are the 'glue' that hold molecules together (Section 2.1.2). During a chemical reaction, some of the bonds in the reactant molecules are broken, and at the same time, new bonds in the product molecules are made. An *input* of energy is needed to *break* a bond between two atoms, the **bond dissociation energy**, but energy is *released* by the *formation* of a new bond. It is the energy released overall from metabolising the molecules in food and drink, which is used or stored to keep the body going.

It is possible to *estimate* the amount of energy released by a chemical reaction by a simple thought experiment. First, balance the equation for the reaction. Second, calculate the energy needed to break every bond in the reactant molecules by adding up all their bond dissociation energies. Third, find the energy released by making all the new bonds in the product molecules. The energy released overall is the difference between these two quantities.

We'll first illustrate the principles involved using the simple reaction of oxygen and hydrogen to make water vapour.

$$2H_2(g) + O_2(g) = 2H_2O(g) \tag{2.16}$$

There are three reactant molecules, two of hydrogen and one of oxygen.

◈ How many bonds in the reactants have to be broken?

◆ Two H—H, and one O=O. (The oxygen molecule has a *double* bond; Section 2.1.4.)

◈ How many bonds are made in the product molecules?

◆ There are two water molecules H—O—H, so four O—H bonds in total.

The bond energy for each type of bond can be obtained from standard tables where the energy values are quoted in kilojoules per *mole* (kJ/mol). The mole is a unit that is a measure of the *same number of molecules* of any compound. Box 2.7 explains why it is important in science.

The energy put into a reaction must have the *opposite sign* from energy given out. The international convention is that energy *put into* a reaction, is positive (+). Energy released or *given out* is given a negative (−) sign. (If you are puzzled about negative numbers, see Box 2.8 on page 46.)

A convenient method for estimating the heat given out by a reaction is to draw up a table with all bonds broken on one side (energy in, *positive*, +) and all bonds formed (energy out, *negative*, −) on the other. For Equation 2.16 this looks like Table 2.1:

It is important to note that in a reaction not all the bonds are actually broken and remade, but for the purposes of the calculation we just assume that they are. The calculation is an *estimate* only because it uses average values for bond energies taken from measurements made on many different molecules. The calculation also ignores factors such as weak bonding *between* molecules.

Table 2.1 Energy chart for Equation 2.16.

type of bond broken	number of bonds broken	bond energy (kJ/mol)	total energy of bonds broken	type of bond formed	number of bonds formed	bond energy (kJ/mol)	total energy of bonds formed
H—H	2	436	+872	O—H	4	464	−1856
O=O	1	498	+498				
Sub total			**+1370**				**−1856**

Box 2.7 (Explanation) Units: what is a mole?

Perhaps without realising it, you will have met the *mole* before, as it is used throughout health science and chemistry and is one of the seven SI base units (Section 1.2). The symbol for the mole is *mol*. For instance, people with diabetes measure their blood glucose levels in *millimoles per litre* (mmol/l, Figure 2.13). Millimoles per litre is also the measure used for the concentration of cholesterol in blood, and you may see it on the labelling of some medicines and vaccines. It is a measure of the *number of molecules* present, rather than a weight.

Using moles of a substance for chemical reactions ensures that the *same* number of molecules are reacting with each other.

Look again at Equation 2.16 and remind yourself what it is telling you: two *molecules* of hydrogen react with one *molecule* of oxygen to produce two *molecules* of water. Each molecule has a mass which is the sum of the masses of its component atoms. The mass of an individual molecule is extremely small, so in the laboratory, scientists scale up to work with a large enough number of molecules to weigh. *A mole is simply a very large number.* The number chosen for the mole (for reasons we won't go into) is called the *Avogadro number*, and it is equal to 6.022×10^{23}.* One mole of hydrogen, H_2 contains 6.022×10^{23} H_2 molecules and weighs 2 g. One mole of oxygen molecules, O_2, which has bigger heavier atoms, weighs 32 g, and a mole of water, H_2O, weighs 18 g.

Follow through this sequence:

Two molecules of hydrogen react with one molecule of oxygen to give two molecules of water:

$$2H_2 + O_2 = 2H_2O$$

Multiplying this equation through by 10 must also be true:

$$20H_2 + 10O_2 = 20H_2O$$

Indeed multiplying through by *any* number must also be true, so, multiplying by the Avogadro number:

$$2 \times 6.022 \times 10^{23}\ H_2 + 6.022 \times 10^{23}\ O_2 = 2 \times 6.022 \times 10^{23}\ H_2O$$

But a simpler way of writing this equation is:

$$2 \text{ moles } H_2 + 1 \text{ mole } O_2 = 2 \text{ moles } H_2O$$
$$(4 \text{ g}) \qquad (32 \text{ g}) \qquad (36 \text{ g})$$

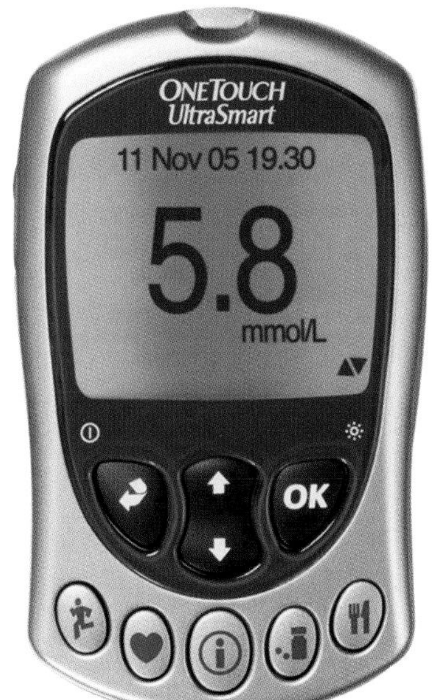

Figure 2.13 A blood-glucose meter calibrated in millimoles per litre (mmol/l). (Photo: Courtesy of Lifescan Inc.)

From Table 2.1, it follows that the overall energy exchange in this reaction is therefore: (+1370−1856) kJ/mol, giving a negative value of −486 kJ/mol. As the result is negative, this means that, overall, energy is *released* or given out by the reaction of hydrogen and oxygen to make water.

*see Appendix, p. 127.

Box 2.8 (Explanation) Negative numbers

Negative numbers work in exactly the same way as an overdraft on a bank statement!

For example, if you have £50 (money in) in your account, but are allowed to withdraw £100 from the cash machine (money out), the final statement will show that you have −£50 in your account. (+£50 − £100 = −£50), i.e. you owe the bank £50!

Now you are in a position to estimate the overall energy released when ethanol is metabolised in the body using the equation you worked out in Section 2.5:

$$C_2H_5OH(g) + 3O_2(g) = 2CO_2(g) + 3H_2O(g) \tag{2.14}$$

It helps to draw out the structural formula of each molecule, so that you can see which bonds are broken and made:

$$\begin{array}{c} \quad\ \ H \quad H \\ \quad\ \ | \quad\ | \\ H-C-C-O-H \quad O{=}O \quad O{=}C{=}O \quad H-O-H \\ \quad\ \ | \quad\ | \\ \quad\ \ H \quad H \end{array}$$

Now try and complete Table 2.2, and calculate the energy of the reaction before looking at the completed table (Table 2.3) and the answer.

Table 2.2 Energy chart for Equation 2.14.

type of bond broken	number of bonds broken	bond energy (kJ/mol)	total energy of bonds broken	type of bond formed	number of bonds formed	bond energy (kJ/mol)	total energy of bonds formed
C−H		413		C=O		770	
O−H		464		O−H		464	
C−C		347					
C−O		358					
O=O		498					
Sub total							

Table 2.3 Completed energy chart for Equation 2.14.

type of bond broken	number of bonds broken	bond energy (kJ/mol)	total energy of bonds broken	type of bond formed	number of bonds formed	bond energy (kJ/mol)	total energy of bonds formed
C—H	5	413	+2065	C=O	4	770	−3080
O—H	1	464	+464	O—H	6	464	−2784
C—C	1	347	+347				
C—O	1	358	+358				
O=O	3	498	+1494				
Sub total			**+4728**				**−5864**

The energy released in this reaction is therefore:
+4728 kJ/mol − 5864 kJ/mol = −1136 kJ/mol.

We can now convert this value from moles into grams: a mole of ethanol weighs 46 g, so the energy released by burning *one* gram of ethanol is (−1136 kJ/mol ÷ 46) = −24.7 kJ per gram (kJ/g).

If a similar calculation for a pure sugar, glucose (formula $C_6H_{12}O_6$) is carried out, the energy released by burning one gram of glucose is −12.9 kJ/g. And you can see that ethanol supplies *almost twice as much energy weight for weight as a pure sugar*. This is topped only by fats which are the most concentrated sources of energy. Table 2.4 lists the approximate energies, in kilojoules per gram, of some other foods for comparison.

Because ethanol releases so much energy when it burns, it is a possible source of renewable fuel (Box 2.9 overleaf).

To return now to the question posed in the title of Section 2.6: 'does drinking alcohol make people fat?' It is a common observation that people who drink alcohol excessively tend to be overweight. Indeed, most diets undertaken for weight loss insist upon the elimination of alcoholic drinks from the diet. But why? The evidence suggests that the body uses a number of different routes to metabolise different foods, and the overall picture is very complex. One popular theory builds on the observation that the two main sources of energy in food, carbohydrates (starch, sugar) and fats, can both be stored in the body when they are not used immediately. This strategy of storing excess food within the body has been highly advantageous throughout evolution, allowing humans to survive through times of famine.

Unlike carbohydrate and fat, ethanol does not have a form in which it can be stored within the body. Thus when ethanol is taken into the body, all the energy it contains has to be used immediately. This means that any carbohydrates and fats taken in at the same time are not needed for energy, so they are stored. So drinking wine with a meal, for example, means that the energy from ethanol will be used immediately by the body, and the energy supplied by food will go straight into storage, for example as fat deposits. However, it also slows the rate of increase in blood sugar after a meal, which may have some long-term health benefits (Brand-Miller et al., 2007).

Scientists often write kJ/mol as kJ mol^{-1} and kJ/g as kJ g^{-1}.

Table 2.4 Approximate food energies.

food	energy (kJ/g)
olive oil	37
butter	32
ethanol	25
milk chocolate	24
cheddar cheese	17
sugar/honey	13
pork sausage	14
salmon	6
chicken	4
cod	3
most fruit and vegetables	1–3

Box 2.9 (Enrichment) Ethanol as an alternative fuel; the production of bioalcohol

The energy value per gram of ethanol that was estimated above is known as an *energy density*. This is a very convenient way of comparing the energies supplied by different fuels. Figure 2.14a shows some comparative values for a selection of commonly used fuels, and you can see that ethanol performs well – almost as well as oil and its products, and they are all far superior to solid fuels.

Of the fuels listed hydrogen looks by far the best – but remember that there is no natural source of hydrogen, and so energy has to be expended in producing it from other chemicals.

Ethanol can be used to fuel cars either alone or as an additive in diesel. Even though ethanol does not perform quite as well as oil, it has the huge advantage that it is a continually renewable source. For this purpose it is currently produced mainly from the fermentation of either corn syrup or sugar cane. By using ethanol in part, Brazil has already eliminated its dependence on foreign supplies of crude oil. Colombia and the USA are also developing large bioalcohol programmes.

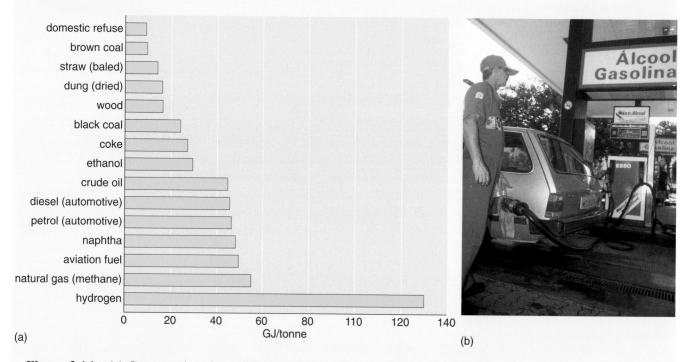

(a)
(b)

Figure 2.14 (a) Comparative chart of fuel energy densities in gigajoules per tonne (GJ/tonne). (Source: data from Internet Encyclopaedia). (b) Alcohol fuel pump in Colombia. (Photo: Joel Creed/Ecoscene)

Not all people who drink alcohol are overweight, however. Many people consciously adjust their intake so that they remain at a stable weight. Some chronic alcoholics, even those appearing to be of a normal weight, are in fact malnourished. They obtain so much of their energy from their drinking that they do not feel the need to eat much food. Consequently, although their energy balance is adequate, they lack the other essential nutrients that food supplies, particularly protein.

Summary of Chapter 2

2.1   Molecules are composed of atoms and can be represented by chemical formulae using chemical symbols.

2.2   The outer electrons of atoms are used in ionic or covalent bonding.

2.3   The surface electron distribution on a molecule affects properties such as its solubility in water and hydrogen bonding.

2.4   Molecules have a three-dimensional shape which can be computer modelled.

2.5   The reactions of molecules can be represented by balanced chemical equations.

2.6   The metabolism of ethanol in the body takes place in the liver by oxidation in stages, producing first acetaldehyde, then acetic acid and finally carbon dioxide and water.

2.7   The conversion of ethanol into acetaldehyde in the liver is facilitated by the enzyme alcohol dehydrogenase (ADH), which acts as a catalyst. Conversion of acetaldehyde into acetic acid in the liver is facilitated by the enzyme aldehyde dehydrogenase (ALDH)

2.8   Different isoforms of ADH and ALDH exist in different individuals and affect the rate of alcohol metabolism.

2.9   The overall metabolism of ethanol in the body produces carbon dioxide and water. From the equivalent chemical equation you can estimate the amount of energy released in the reaction by summing the energies required to break all the bonds of the reactants and the energies released in making new bonds in the products.

Learning outcomes for Chapter 2

After studying this chapter and its associated activities, you should be able to:

LO 2.1   Define and use in context, or recognise definitions and applications of, each of the terms printed in **bold** in the text. (Questions 2.1 to 2.4)

LO 2.2   Describe the bonding and shapes of simple molecules and in particular ethanol, using structural formulae and computer models. Distinguish ionic and covalent bonding, and single and double bonds. (Questions 2.1 to 2.5, and DVD Activities 2.1 and 2.2)

LO 2.3   Describe hydrogen bonding and the process whereby substances dissolve in water. (Question 2.6 and 2.7 and DVD Activities 2.1 and 2.2)

LO 2.4   Use chemical notation correctly to write and balance simple chemical equations, particularly involving the oxidation of ethanol and its metabolism in the body. (Questions 2.8 and 2.9)

LO 2.5   From a chemical equation, be able to estimate the amount of energy released in the reaction from the breaking of all the bonds of the reactants and reforming new bonds in the products. (Question 2.10)

Self-assessment questions for Chapter 2

You have had the opportunity to demonstrate LOs 2.2 and 2.3 by answering questions at the end of DVD Activity 2.2.

Question 2.1 (LOs 2.1 and 2.2)

Nitrogen uses three of its outer electrons for bonding to other atoms. Describe the bonding (how many covalent bonds and what type) in the nitrogen molecule, N_2, and in ammonia, NH_3.

Question 2.2 (LOs 2.1 and 2.2)

When calcium, Ca, forms a positive ion, it loses *two* electrons. What is the charge on the calcium ion? How is this written? In calcium chloride, how many chloride ions, Cl^-, are needed to balance the charges on the calcium ion? What is the formula of calcium chloride?

Question 2.3 (LOs 2.1 and 2.2)

Sulfur dioxide, SO_2, is used for sterilisation in wine-making. Sulfur uses four outer electrons to bond to oxygen in this molecule. Write out a likely structural formula. (Note: SO_2 is a 'bent' molecule.)

Question 2.4 (LOs 2.1 and 2.2)

Incorrectly bonded structural formulae of acetaldehyde and acetic acid are given below. Copy them and correct the bonding by changing single to double bonds where necessary.

$$
\begin{array}{ccc}
\begin{array}{c}
\text{H} \quad\;\; \text{O} \\
| \quad\; / \\
\text{H—C—C} \\
| \quad\;\; \backslash \\
\text{H} \quad\;\; \text{H}
\end{array}
& \text{and} &
\begin{array}{c}
\text{H} \quad\;\; \text{O} \\
| \quad\; / \\
\text{H—C—C} \\
| \quad\;\; \backslash \\
\text{H} \quad\; \text{O—H}
\end{array}
\end{array}
$$

Question 2.5 (LOs 2.1 and 2.2)

Write down what you think are the different uses of the ball-and-stick and the space-filling representations of molecular computer models.

Question 2.6 (LO 2.3)

Describe what happens when potassium bromide, KBr, an ionic compound containing the bromide ion Br^-, dissolves in water (make reference to the partial charges on the water molecule).

Question 2.7 (LOs 2.1 and 2.3)

Explain what the superscripts, subscripts, prefix numbers and the letters in brackets mean in the following equation, where calcium iodide dissolves to form positive calcium ions and negative iodide ions:

$$CaI_2(s) = Ca^{2+}(aq) + 2I^-(aq) \tag{2.17}$$

Question 2.8 (LO 2.4)

Write balanced equations for the following reactions, remembering to include the physical states:

(a) Hydrogen gas with chlorine gas to form gaseous hydrogen chloride, HCl.

(b) Hydrogen gas with nitrogen gas, N_2, to form gaseous ammonia, NH_3.

(c) Dissolving solid calcium chloride, $CaCl_2$ (form the aqueous ions).

Question 2.9 (LO 2.4)

Write balanced equations for the following reactions, remembering to include the physical states:

(a) The combustion of butane gas, C_4H_{10} in oxygen to form carbon dioxide and water vapour – it helps here to start with two molecules of butane.

(b) The fermentation of sugars is used to prepare ethanol for human consumption. This is carried out in the presence of yeast, but in the absence of oxygen. Carbon dioxide is also produced. Balance the equation for this reaction using glucose, $C_6H_{12}O_6$, as the sugar – and it helps to know that two molecules of carbon dioxide are produced.

Question 2.10 (LOs 2.1 and 2.5)

Calculate the energy given out by the burning of methane, CH_4. (Hint: Look back to Equation 2.10.) Bond energies are given in Table 2.2.

CHAPTER 3

WHERE DOES ALCOHOL GO IN THE BODY?

This section follows the journey of the ethanol molecule through the body to the organs.

3.1 Absorption of alcohol from the gut

When an alcoholic drink, or any other substance, is swallowed it is often considered to have *entered* the body. However, this is not necessarily the case. The gut (Figure 3.1) is essentially a long tube running from the mouth to the anus. Food that is swallowed is passed along inside this tube by ripples in the muscular wall (a process known as peristalsis; perry-stahl-sis). The tube has various distinct regions (e.g. the mouth, oesophagus, stomach, small intestine, large intestine) whose shapes and dimensions enable substances that have been taken in (*ingested*) to be broken down as they pass along it so that they can be absorbed into the bloodstream. The **absorption** process involves the molecules

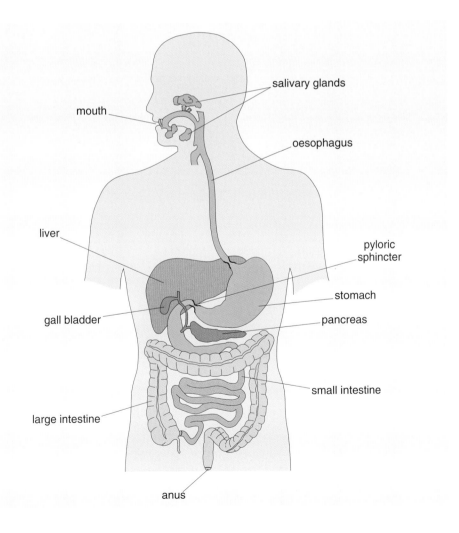

In life, the liver lies over and around the stomach; here it is shown behind to allow the anatomy of the gut to be seen.

Figure 3.1 Schematic diagram of the gut and associated organs. The gut is a continuous tube that runs all the way from the mouth to the anus, forming the oesophagus, stomach, small intestine and large intestine.

that are released from the digested food passing through the wall of the gut and into the surrounding blood vessels – at this point the food or drink can be considered to have truly 'entered' the body. Food that is not absorbed, such as plant fibre, passes straight through the gut without leaving the inside of the tube, i.e. not entering the body tissues, and is expelled (*egested*) as waste.

When food or drinks are swallowed, they pass through the oesophagus into the stomach, which is a bag-like structure where food is mixed with digestive enzymes, before passing into the small intestine through the pyloric sphincter. The **pyloric sphincter** is a muscular constriction which closes when food is present in the stomach (Figure 3.2), ensuring that food is sufficiently digested before it is released into the small intestine. Ethanol is absorbed into the blood at a greater rate through the wall of the small intestine compared with the wall of the stomach (because of the large surface area of the intestine which is discussed below). Therefore if ethanol is retained in the stomach (by the closure of the pyloric sphincter) it will be absorbed more slowly, but once the ethanol passes into the intestine it will be absorbed rapidly.

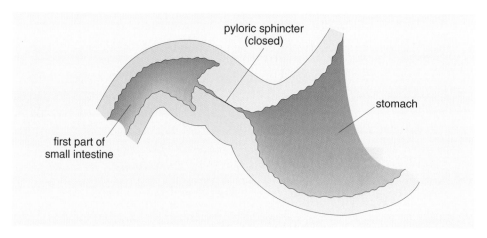

Figure 3.2   The pyloric sphincter.

◆ If there is food in the stomach, what effect will it have on the speed with which the ethanol from a drink enters the bloodstream?

◆ The presence of food in the stomach will delay the rate at which ethanol in the stomach enters the small intestine because the pyloric sphincter will be closed to retain the food (Figure 3.2).

On an empty stomach the concentration of ethanol in the blood reaches a peak about an hour after consumption, because it passes straight through the pyloric sphincter and is absorbed rapidly in the small intestine (look back to Figure 2.10).

You do not need to know the details of the digestive system, but it is useful to understand that food is mechanically ground up by the teeth, then digestive enzymes break it down chemically in the mouth, stomach and intestine (Figure 3.3). Absorption of nutrients from broken-down food takes place in the small intestine, and water is removed from the waste in the large intestine before it is egested.

Figure 3.3   Diagram illustrating the processes occurring in the different parts of the gut.

food

mouth

oesophagus

stomach

food broken down by enzymes

small intestine

absorption of nutrients

large intestine

removal of water

anus

The structure of the small intestine is highly specialised to facilitate the absorption of substances from the gut into the bloodstream. The small intestine is a tube approximately 5 m long, with a diameter of about 4 cm. The surface area of the lining of this tube is greatly increased by the presence of numerous *villi* (vill-eye, singular, *villus*) – finger-like projections about 1 mm long which extend into the lumen (Figure 3.4a, b).

The cells which form the surface of the intestinal lining are of the type known as *epithelial* (ep-ee-thee-lee-al) *cells* and they have projections on the surface facing into the gut called microvilli which further increase the surface area. The result is that the lumen of the human small intestine has a huge surface area of about 300 square metres (300 m^2) – about the same area as a tennis court.

◈ What effect will this enlarged surface area have on the ability of the small intestine to absorb nutrients?

◆ Absorption of nutrients into the blood takes place through the wall of the small intestine, so a larger surface area means absorption of nutrients can occur at a faster rate.

> Lumen is the name biologists give to the region inside a tube.

> Epithelial cells usually form a barrier (such as the skin) or an interface across which substances are absorbed or secreted (as in the digestive system).

> Lymph vessels are part of the body's natural drainage system, the lymphatic system. The lymphatic system is discussed in another book in this series *Screening for Breast Cancer* (Parvin, 2007).

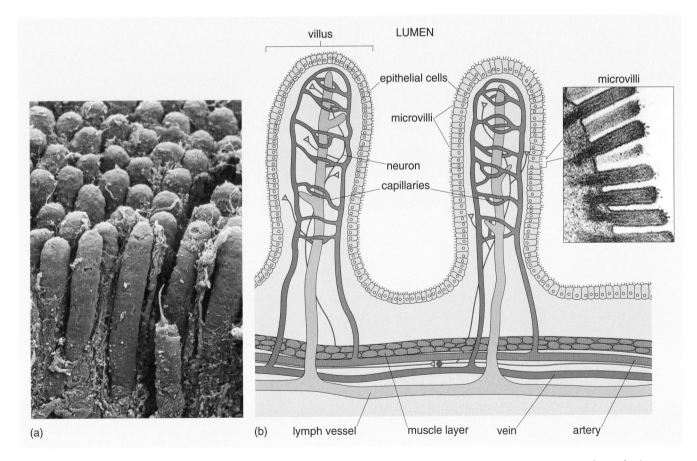

Figure 3.4 (a) Electron micrograph and (b) diagram of villi, regular finger-shaped projections, approximately 1 mm in length, of the wall of the small intestine which increase its surface area. Inset to (b) Microvilli on the epithelial cells increase the surface area still further. The wall of the small intestine contains layers of muscle (this enables movement of gut contents by peristalsis) and many blood and lymph vessels. (Photos: Science Photo Library)

Absorption of ethanol takes place via a vitally important process called *diffusion* which occurs throughout the body (see Box 3.1).

Box 3.1 (Explanation) Passive diffusion

The molecules in a liquid or gas are constantly on the move, bouncing around randomly in between collisions with other molecules. They therefore tend to spread out to fill any space that is available. **Diffusion** is the *net* movement of molecules from regions of high concentration into regions of lower concentration, until there is an even distribution throughout the available volume. Think about a drop of dye in a glass of water. Even if you don't stir it, after several hours the dye will have spread evenly throughout the water (Figure 3.5a). The same principle applies to movement of molecules across a biological membrane that is *permeable* to those molecules (that is, it cannot act as a barrier against them). Biological membranes are usually permeable to very small molecules and ions such as water, oxygen, carbon dioxide and dissolved salts. Movement of these molecules and ions across the membrane occurs by *passive* diffusion (that means no energy expenditure is required) as long as the concentration is greater on one side of the membrane than it is on the other (as in Figure 3.5b). In this situation a *concentration gradient* is said to exist. You could think of molecules 'rolling down' the gradient from the high side to the low side. Unless other forces oppose them, molecules will always diffuse throughout the available space, like the drop of dye, until the concentration gradient is abolished.

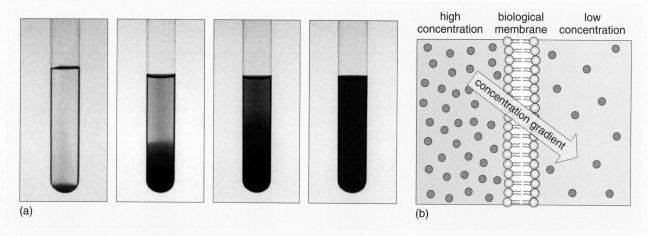

Figure 3.5 (a) Diffusion of purple dye in a test tube containing water (i) shortly after setting up; (ii) after 7 hours; (iii) after 24 hours (iv) after 3 days. (b) Schematic representation of a concentration gradient across a biological membrane, showing the direction in which dissolved molecules diffuse until the concentration equalises on both sides of the membrane.

Molecules of ethanol are not broken down in the gut, but diffuse through the wall of the gut into the adjacent blood vessels. After an alcoholic drink the concentration of ethanol in the small intestine is greater than the concentration in the blood, so diffusion results in a net movement of ethanol molecules out of the gut lumen and into the blood.

Blood is transported around the body in pipes known as blood vessels. Large vessels that carry blood away from the heart are known as **arteries** and large vessels that carry blood towards the heart are **veins**. The arteries and veins branch into a network of tiny **capillaries** that infiltrate all of the tissues of the body (Figure 3.4).

The circulatory system is described briefly in Activity 3.1, and in more detail in other books in this series, Midgley (2008) and *Trauma, Repair and Recovery* (Phillips, 2008).

Activity 3.1 The ethanol journey

Allow 20 minutes

Now would be the ideal time to study the DVD activity entitled 'The ethanol journey' on the DVD associated with this book. If you are unable to study it now, continue with the rest of the chapter and return to it as soon as you can.

An animated sequence shows the journey of ethanol molecules from an alcoholic drink as they pass down the gut and are absorbed in the small intestine. The absorption process is highlighted with an explanation of diffusion of ethanol from the gut lumen into the bloodstream. Following absorption, the journey continues via the portal vein to the liver, then on to the other organs, until all of the ethanol is removed from the circulation.

3.2 Transport of absorbed ethanol in the body

The small blood vessels that carry blood away from the gut converge to form a large vein, the *portal vein*, which leads directly to the liver where some of the ethanol is removed (Figure 3.6). Blood that leaves the liver via the hepatic vein contains the remainder of the dissolved ethanol and enters the general circulation, carrying ethanol to all of the organs and tissues in the body.

Hepatic means 'of the liver'.

The blood vessels themselves are affected by the presence of ethanol in the blood. Following consumption of an alcoholic drink, the drinker's skin becomes warmer and flushed due to an effect called **vasodilation** (vayzo-dye-lay-shun) during which the narrow blood vessels immediately beneath the skin *dilate* – they get wider so they carry more blood. This mechanism, together with sweating, is normally used as a way of cooling the body when it is too hot – by increasing the

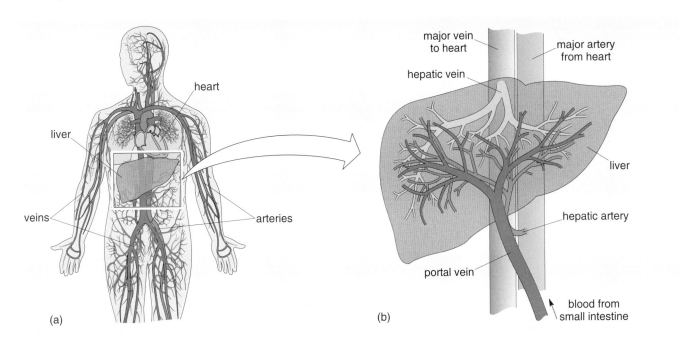

Figure 3.6 (a) Position and (b) blood supply of the liver. Notice that blood flowing into the liver comes from both the hepatic artery and from the portal vein. Blood leaves the liver through the hepatic vein.

flow of blood just under the surface of the skin, more cooling can occur, making this a good way for excess heat to escape. This effect occurs when alcohol is consumed, even in cold conditions, and is potentially dangerous since excessive heat loss from the body could result in *hypothermia* where the core organs become too cold to function properly.

◆ Is an alcoholic drink a good way to warm someone up on a cold evening?

◆ An alcoholic drink might indeed make a cold person *feel* warmer as their skin will warm up due to vasodilation, but it would *increase* their actual loss of body heat to the environment, so their core body temperature would *fall*.

As ethanol circulates throughout the body dissolved in the bloodstream, it is able to diffuse into cells in all of the organs in the body. The rate at which ethanol leaves the blood and enters the organs is dependent upon how rich the blood supply is to the particular organ. Organs with a particularly rich blood supply include the brain and the lungs, so ethanol will tend to affect these organs sooner than others. Because it is able to diffuse freely, ethanol becomes distributed throughout the water-based components of the body, that is, the blood and the cells of most tissues such as muscles and brain. However, very little ethanol diffuses into fatty tissue because ethanol is more soluble in water than it is in fat (Section 2.3.1). On average, women have a higher proportion of body fat than men of the same weight, therefore the relative volume into which ethanol can freely diffuse is smaller in women than in men of equivalent weight.

◆ If a man and a woman of identical weight each simultaneously drank two glasses of wine, what would you expect to observe if you measured their blood-ethanol concentration 1 hour later and why?

◆ The blood-ethanol concentration would be higher in the woman than the man. The woman has a smaller volume of water-based tissues for the ethanol to dissolve in than the man because a higher proportion of her body weight is fat.

The concentration of ethanol in the blood is known as the blood-alcohol concentration, BAC (Section 2.4.1).

3.3   Ethanol from the gut first passes through the liver

The liver plays an important role in a number of body functions, including digestive processes, regulation of blood glucose levels, storage of vitamins and the destruction of noxious substances (of which ethanol is one example).

The liver receives three-quarters of its blood supply from the portal vein, carrying blood directly from the gut, and only one-quarter from the general circulation (oxygen-rich blood pumped from the heart via the hepatic artery, Figure 3.6). All of the blood leaving the liver via the hepatic vein drains into the general circulation. This means that all of the blood from the portal vein, complete with any absorbed nutrients or other substances (ethanol, toxins, drugs, etc.) passes through the liver before reaching the general circulation.

◆ What advantage and what risks can you see in this arrangement of the blood flow to the liver?

◆ While this arrangement means that the liver can efficiently process absorbed substances such as ethanol and other toxins, it also means that the liver is exposed to higher concentrations of ethanol than any other organ.

This increased exposure of the liver to ethanol relative to the other organs plays an important role in the damage processes that excessive ethanol consumption can cause (Chapter 5).

3.4 The effect of ethanol on the kidneys

The kidneys are responsible for **excretion** of waste substances from the blood through the production of urine. Excretion specifically means the *separation* of waste products from the blood, as distinct from egestion which refers to the *ejection* of waste from the digestive system (Section 3.1). Drinking alcoholic beverages has a familiar and characteristic effect on this process – the kidneys excrete *more* urine than would be predicted from the volume of liquid consumed, a phenomenon familiar to all drinkers! In order to understand this phenomenon it is important to know the general principles of kidney function.

There are two kidneys in the human body, situated at the back of the abdomen just below the ribs. These fist-sized organs are responsible for removing excess fluid, minerals and waste products from the blood by producing urine for excretion. Each kidney contains around a million structures known as **nephrons** which filter the blood and produce the urine (Figure 3.7).

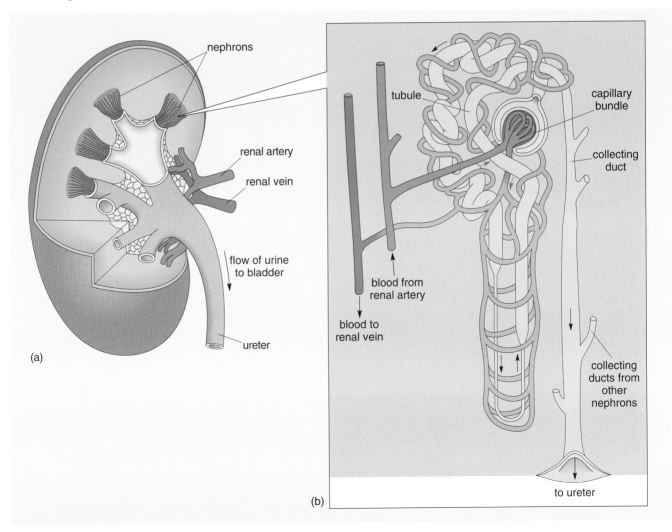

Figure 3.7 (a) Diagrammatic representation of the kidney structure. (b) An enlarged diagram of a nephron.

Renal means 'of the kidneys'.

Vasopressin (sometimes called *anti-diuretic hormone*) is introduced in another book in this series, *Pain* (Toates, 2007).

Blood enters the kidney via the renal artery which branches into many thinner blood vessels, eventually becoming capillaries, the thinnest blood vessels of all (Section 3.1). Capillaries have thin walls through which fluid can leak. In the kidney these thin-walled vessels form tight bundles which contain blood at high pressure. Fluid leaks from the blood under pressure through the walls of the capillaries and is collected and fed into the tubule of the nephron. Because large molecules and blood cells remain in the capillaries and only the fluid part of the blood and some dissolved smaller molecules pass into the tubule, this is a *filtration* stage. As the filtered liquid passes along the winding tubule of the nephron, various valuable components that the body requires (glucose, ions and most of the water) are taken back into the blood, leaving only the unwanted waste materials and remaining water, to drain into the collecting duct and form urine.

Controlling the amount of water and other molecules that are excreted by the kidneys is a critical process for maintaining the body in a functional state. One important mechanism by which kidney function is regulated is by a hormone called *vasopressin* (vayzo-press-in). **Hormones** are chemical signalling molecules that circulate in the blood and trigger responses in specific tissues and organs in the body. Vasopressin is secreted by the brain in response to detection of decreased water levels in the body. This hormone travels in the blood to the kidneys, where it increases the ability of the cells which line the nephron tubules to move more water from within the tubules back into the bloodstream. The overall effect when vasopressin is released by the brain is a reduction in the amount of water that is lost into the urine.

◆ What effect do you think ethanol has on this hormone system (recall, we said earlier that ethanol makes the kidneys excrete more urine than the volume of liquid consumed)?

◆ Ethanol inhibits the release of vasopressin, thus causing an increase in the amount of water that is lost in the urine. In turn this disproportionate loss of water from the body causes dehydration, and a thirst for more drinks!

The effect of ethanol on the kidneys via vasopressin is an indirect one: the ethanol acts on the brain to modify the release of a hormone, which then travels via the bloodstream to act on the kidneys. The regulation of the activity of distant organs through the release of hormones is part of the **endocrine** (end-oh-krin) system, an important signalling pathway within the body.

Summary of Chapter 3

3.1 Ethanol is absorbed from the gut into the bloodstream by passive diffusion down a concentration gradient, which does not require the expenditure of energy.

3.2 Most ethanol absorption takes place in the small intestine which has a large surface area for the absorption of digested food. Movement of ethanol from the stomach to the small intestine (and thus absorption) is delayed when food is present in the stomach, due to the closure of the pyloric sphincter.

3.3 Absorbed ethanol travels from the gut via the portal vein to the liver where some is removed before the remainder is transported around the body in the general circulation.

3.3 Ethanol becomes distributed throughout the watery tissues of the body but little is absorbed by fatty tissues. Women have a higher proportion of body fat and so tend to exhibit a higher blood-alcohol concentration (BAC) than men of the same weight with similar levels of alcohol consumption.

3.4 Ethanol acts on the brain, inhibiting the release of vasopressin and thus allowing the kidneys to make more urine.

Learning outcomes for Chapter 3

After studying this chapter and its associated activities, you should be able to:

LO 3.1 Define and use in context, or recognise definitions and applications of, each of the terms printed in **bold** in the text. (Questions 3.1, 3.2 and 3.4)

LO 3.2 Describe the absorption of ethanol from the gut, with reference to the role of the pyloric sphincter and the process of diffusion. (Question 3.1 and DVD Activity 3.1)

LO 3.3 Describe how ethanol travels around the body dissolved in blood, and how its distribution within organs and tissues is related to their blood supply and fat content (male/female differences). (Question 3.3 and DVD Activity 3.1)

LO 3.4 Explain why the liver is exposed to higher blood-ethanol concentrations than other organs. (Question 3.3 and DVD Activity 3.1)

LO 3.5 Describe the effect of ethanol on blood vessels and on kidney function. (Questions 3.2 and 3.4)

Self-assessment questions for Chapter 3

Question 3.1 (LOs 3.1 and 3.2)

What effects on intoxication does drinking on an empty stomach have and why?

Question 3.2 (LOs 3.1 and 3.5)

Explain why drinking alcohol can make people's faces turn red.

Question 3.3 (LOs 3.3 and 3.4)

Chapter 1 identified cirrhosis of the liver as a major cause of illness and premature mortality among people who drink excessive amounts of alcohol over a long period. Explain why this organ may be affected more than others.

Question 3.4 (LOs 3.1 and 3.5)

Why does drinking alcohol result in the production of more urine than when an equivalent volume of water is consumed?

CHAPTER 4

ALCOHOL – THE LINKS TO BRAIN, BEHAVIOUR AND MIND

4.1 How to understand why people drink alcohol

Alcohol is well-known for its effect on people's state of mind and behaviour. Several familiar examples illustrate this. The ability to perform skilled activities, such as driving, is disrupted by alcohol. Speech tends to become slurred. While under alcohol's influence, there can also be an increased risk of violence, accidents and socially 'inappropriate' behaviour (Chapter 1). There is the possibility that its intake will become damaging to mental health and, in the extreme, addictive. On a more positive note, alcohol tends to increase people's sociability. Thus, taken in moderation and in an appropriate setting, alcohol can have beneficial effects.

This chapter investigates the reasons why people drink alcohol and the effects that alcohol has on brain, behaviour and mind: it explores both the science of brain processes – neuroscience – and the resulting social behaviour. A logical starting point for the study of alcohol is to consider the 'human zoo'.

4.1.1 An evolutionary perspective

It is sometimes said that these days many people live in a 'human zoo': in other words, like animals housed in a zoo, most people live in an environment that is very different from that of their hunter-gatherer ancestors. To understand features of behaviour that are seen today, it can be useful to speculate on what life was like in the past. Although alcohol has been consumed by humans for thousands of years, an abundant supply of alcohol appears to be very recent. Therefore, to understand the effects of alcohol in terms of evolution, it is necessary to consider *natural rewards*. A **reward** is something that animals, including humans, strive to obtain and will invest effort in gaining. A 'natural reward' refers to such things as food, water and sex, as distinct from such 'unnatural rewards' as alcohol. The biological advantage of seeking natural rewards is obvious: our survival throughout evolution has depended upon brain mechanisms that have caused us to eat, drink and mate.

The notion of the 'human zoo' is introduced in another book in this series *Water and Health in an Overcrowded World* (Halliday and Davey, 2007).

So-called 'recreational drugs', such as alcohol, exhibit some similar properties to natural rewards in terms of the desire to obtain them, some of their effects on the brain, and the tension created when access is denied. Recreational drugs affect some of the same brain regions as do natural rewards. Indeed, the evidence suggests that the attraction of drugs arises from their capacity to influence the brain processes that have evolved for dealing with natural rewards (Nesse and Berridge, 1997). Their effects can sometimes be even stronger than those of natural rewards.

4.1.2 A psychobiological perspective

To jump from the evolutionary past to the present, the chapter will first take a psychological perspective on the causes of drinking and the effects of alcohol. It will then relate this perspective to the effects of ethanol on the brain. Finally, it considers how moderate 'social use' can tip into excessive intake.

In relating psychology and biology, the chapter will adopt a **psychobiological perspective**, making the following two related assumptions:

- Both psychology and biology are necessary for understanding behaviour (e.g. seeking alcohol) and the effects that chemicals have on the mind and behaviour.

- Changes in the activity of the *brain* (for example, as a result of ethanol entering the brain) are associated with corresponding changes to the *mind* and to *mental* experience, as well as changes in behaviour.

The topic of mind and brain is described in another book in this series (Toates, 2007).

Stated in other words, understanding brain and mind is guided by the assumption that the mind is 'an expression of the activity of the brain'. For example, changes in the electrical activity within the brain would be said to be the biological *basis* of any changes in mood. Thus, the difference between joy and despair is characterised by differences in the activity of certain regions of the brain. Interventions, such as taking a medicinal drug for anxiety or a depressed mood, or taking alcohol, act because the drug affects the brain and changes its activity. Thereby, correspondingly, mental states, including mood, are altered.

The activity of the **neurons** of the brain is affected by molecules of ethanol arriving in the blood. Neurons are a type of cell found throughout the nervous system and are involved in communication and processing of information. By changing the activity of neurons, there are associated changes in mind and behaviour.

This chapter investigates the factors that determine behaviour and our mental states and asks how these factors are altered by alcohol.

4.2 Psychological perspectives on drinking alcohol

As shown in Figure 4.1, intake of any alcohol is determined by what is termed the 'brain and mind'. Reciprocally, alcohol affects the brain and mind. The reasons people drink alcohol are due, of course, in part to its past effects on the body, so cause and effect are inextricably linked.

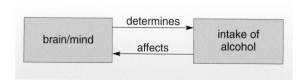

Figure 4.1 The brain/mind determines intake of alcohol and is affected by it.

4.2.1 Causes of drinking

There are various external and internal events that cause alcohol to be drunk. The combination of factors includes our emotions, such as fear or frustration, stimuli present in the immediate environment (e.g. the sight of someone else drinking), memories of past events and intentions for the immediate future (e.g. to lift anxiety). Of course, associated with these psychological causes are social factors, such as a wish to be 'one of the group', unemployment, social competition or alienation from mainstream society.

A powerful factor that plays a role in the tendency to drink alcohol is **classical conditioning** (Box 4.1).

Box 4.1 (Explanation) Classical conditioning

The best-known example of classical conditioning is Pavlov's demonstration of the response of salivation in dogs. All hungry dogs produce saliva ('salivation') as the response to food in the mouth. Of course, the sound of a bell does not normally trigger salivation. On a number of occasions, Pavlov sounded a bell at the same time that food was presented to the dog. As a result of this repeated pairing, the sound of the bell on its own acquired the ability to trigger salivation (Figure 4.2). The sound became known as a *conditional stimulus*, as its power to trigger salivation was conditional upon its earlier association with food. The term 'classical conditioning' reflects the fact that this was the first form of conditioning to be demonstrated. In honour of its discoverer, it is also known as 'Pavlovian conditioning'.

If the bell was sounded repeatedly on its own (without the food) it lost its power to trigger salivation. This is termed **extinction**.

"AND THEN INSTEAD OF FEEDING ME HE WOULD RING A LITTLE BELL."

Figure 4.2 Pavlov's dog. (Source: www.CartoonStock.com)

In people with a history of drinking alcohol, classical conditioning comes to play an important role in triggering alcohol ingestion, since drinking has often been regularly associated with certain conditional stimuli (Weiss, 2005). These stimuli trigger the desire for alcohol.

◆ Can you suggest some cues that are often associated with drinking alcohol and thereby become conditional stimuli with the ability to trigger the desire for alcohol?

◆ Such things as being near a bar, the sight of an off-licence or a regular drinking companion, advertisements for alcohol, and, in the case of combined smokers/drinkers, smoking a cigarette.

Figure 4.3 The context of drinking.
(Photo: Dawn Partner)

Classical conditioning has been suggested as a possible factor in explaining binge drinking (Tomie, 1996). An empty glass in the hand, particularly a spirit, wine or beer glass, is a conditional stimulus associated with alcohol, as is the environment itself (Figure 4.3), all of which can trigger alcohol consumption.

Consider another factor as follows. We can all recall events of many years ago, and also imagine future events. We thus estimate the likely consequences of our actions such as drinking alcohol, and base our behaviour, in part, on these estimated outcomes.

The decision on whether to take a drink, or to decline it, is influenced by many interacting factors. Some of these are represented by stimuli physically present (e.g. the half-empty bottle on the table) and others are based on estimating the future benefits or harm associated with alcohol. It is characteristic of their behaviour, that heavy drinkers often discount such factors as future harm (think back to Activity 1.3 and the accounts of risk-taking by the students).

4.2.2 Effects of alcohol

Ethanol from alcoholic drinks is taken into the body and absorbed into the bloodstream and hence into the brain. By entering the brain, ethanol affects the activity of neurons and hence the mind. The brain can be described as performing **information processing** (e.g. calculating the width of a space in which to park a car) and this is, in general, disrupted by ethanol. This section will look at some effects of alcohol, both intended and unintended by the drinker.

Information processing is described in another book in this series (Toates, 2007).

Table 1.6 (Chapter 1) showed the influence of alcohol on traffic accidents. What could a psychological perspective contribute to understanding here? Imagine a person is driving a car whilst engaged in animated conversation with a passenger. It is a familiar route and the bends are negotiated in a fairly automatic way. Not much attention is demanded by the driving and attention is focused on the conversation – all is going well. Then a car pulls out unexpectedly from a side road and a collision is probable. Attention needs to be shifted from the conversation to the task of driving. Alcohol slows the speed of reaction in such a situation. The link between sensory event (seeing the car) and motor (muscular) reaction (putting foot on brake) is delayed. Psychologists have carried out experiments to see more precisely what exactly causes the slowing of reaction, and we now go on to describe these.

How do we know for sure that alcohol disrupts driving skills? To demonstrate this, psychologists create laboratory 'simulations' of driving tasks. These bear some similarity to the task of driving but the measures of performance can be made accurately and do not involve risk to either psychologist or participant. For example, in one study, people were asked to perform a visual task – decide which of two lines is the longer. The influence of alcohol on speed and accuracy in performing the task is measured. Another task is described as 'choice reaction-time'. This might consist of two keys that need to be pressed, one to the left and one to the right. If a green light shines, the one to the left needs to be pressed as rapidly as possible, whereas a red light is the cue to press the key to the right. All such tasks are slowed by alcohol (Figure 4.4).

Other tasks are exemplified by walking upright along a white line on the floor. Alcohol increases the errors in doing this. This exemplifies ethanol's role in disrupting the brain's ability to coordinate between sensory input (the sight of the white line) and motor output (control of the skeletal muscles of the legs). The brain monitors the position of the body and, if, for example, the body starts to sway, the brain automatically causes corrective action to be taken. That is, the brain alters the signal to the muscles to compensate. If the underlying processes are affected by ethanol this compensation is disrupted, explaining part of its contribution to accidents.

Ethanol has direct chemical effects on the neurons of the brain. However, another factor underlying the effect of alcohol is the context in which it is taken. For example, giving someone a neutral liquid, a *placebo* (plah-see-boh), which that person *believes* to be alcoholic will sometimes trigger some of the same effects on their brain, mind and behaviour that *real* alcohol triggers. This is termed a **placebo effect**. Experimenters who wish to see the chemical effects of alcohol uncomplicated by the placebo effect need to be very careful in how they design their experiments. For instance, the alcohol can be disguised by a strong taste that masks the taste of alcohol, and a control drink with the same masking taste but not containing alcohol can be given and the effects of the two drinks compared; in one of the experiments which you will see shortly in Activity 4.1b on the DVD, the control drink is tonic water, flavoured with angostura bitters, and the alcoholic drink is the same with portions of pure (unflavoured) ethanol added.

People are very bad at monitoring their level of intoxication and often underestimate just how much they have drunk (Section 1.2). They also tend to underestimate the effect that a given level of ethanol will have on their ability to perform skilled behaviour such as driving. For example, in driving simulation tests conducted on bus drivers (Paton, 2005), it was shown that, with blood-ethanol levels of 50 mg per 100 ml, they believed that they could drive through gaps that were, in reality, too narrow for their vehicles. Presumably, most people who drink alcohol do so *in spite of* the fact that it impairs skilled behaviour. Rather, they drink because of the mood-altering effects of alcohol.

Alcohol reduces anxiety over a wide range of circumstances, though not all (Josephs and Steele, 1990). Drinkers typically learn an association between drinking alcohol and a reduction in anxiety experienced shortly afterwards. Psychologists believe that this is a powerful basis of its effect as a reward, or, as it is sometimes termed, a **reinforcer**. In psychology, a reinforcer refers to a factor that strengthens a tendency to engage in a particular behaviour. In the present case, the tendency to drink alcohol is increased by its anxiety-lowering effects in the past. Alcohol can lower painful self-consciousness, hence improving one's own self-image (Hull et al., 1986).

Figure 4.4   Equipment for testing choice reaction-time using two letters. (Photo: Courtesy of Neuro-Test Inc.)

Chapter 1 quoted George Bernard Shaw: 'Alcohol is the anaesthesia by which we endure the operation of life'. Figure 1.7 showed the reasons people drink alcohol, in large part to self-medicate.

4.2.3 Inhibition and alcohol myopia

A fundamental change caused by ethanol is the lifting of certain inhibitions on behaviour, under particular conditions. This is the basis of its reputation as a 'social lubricant'. Even at low doses, ethanol tends to have the effects of increasing sociability and confidence, while making mood more positive. Under its influence, people tend to talk more (Koob and Le Moal, 2006). The caution expressed by 'tend to' highlights that these effects do not always occur. Ethanol can have very variable effects, even within a single individual. It can sometimes lift one's mood and yet at other times make depression worse.

Chapter 1 noted the association between alcohol and violence, and psychological insights are useful here. A lifting of inhibition can explain how ethanol can facilitate violence in some individuals. This might be manifest as physical violence or just aggressive remarks, and it is associated with underlying anger and frustration. These tendencies would normally be held in check by respect for social conventions and not wanting to incur social disapproval, retaliation and punishment by the law. Alcohol often lowers such restraints.

You first met the notion of alcohol myopia in DVD Activity 1.3, where the student group described their experiences of this phenomenon.

Alcohol tends to produce what is appropriately termed **alcohol myopia** (Steele and Josephs, 1988). Myopia means literally 'short-sightedness'. By analogy, the term 'alcohol myopia' indicates alcohol's effect in inducing 'psychological short-sightedness'. Alcohol lowers the range of attention. It gives a bias in favour of attending to *immediate* internal and external cues and against considering more remote events. For example, suppose that (a) you are choosing paint colours for decorating your home and (b) you must take an exam next week and you are feeling anxious about it. The exam is still a distant event. Alcohol will tend to make it less likely that the focus of attention will shift to the distant task, it being fully occupied with the immediate demands of the task of colour choosing. This will lower the anxiety that attending to exam revision would otherwise trigger.

Alcohol does not invariably lower anxiety. It appears to do so in the situation where a person is engaged in an activity that demands some attention, such as conversation. The solitary drinker with nothing else to do but to drink might even find his or her anxiety *increased* by alcohol. This is summed up by the common term 'crying in one's beer' (Steele and Josephs, 1988). Hence, the effects of alcohol can only be understood in terms of its chemical properties *in interaction with* the cues in the environment and our expectations and memories.

For another example, the risk of an accident when driving is usually not immediately apparent. Alcohol will tend to divert attention from this risk in favour of the immediate concern of getting from one place to the next. This is in addition to its effect in disrupting sensory-motor coordination.

Alcohol myopia might have something to contribute to understanding the reason why alcohol is associated with increased tendencies to suicide (Chapter 1). Alcohol makes people more impulsive, i.e. acting on the 'here-and-now', while tending to disregard the future.

A controlled experiment demonstrated alcohol myopia (Zeichner and Phil, 1979; see Figure 4.5). Participants in group A were subject to painful stimulation (an unpleasant sound delivered through ear-phones) which they were told was

(a)   (b)   (c)

Figure 4.5   Experiment to demonstrate alcohol myopia: (a) and (b) in the absence of alcohol and (c) after drinking alcohol. (a) Group A participant receives aversive stimulation; (b) group A participant imagines he is retaliating against a group B participant; (c) the situation after alcohol is drunk.

administered by a fellow participant in group B. In fact, there was no group B, only a computer. The sound could be stopped by participants in group A retaliating with an unpleasant sound to the imagined person in group B. The group A participants selected the intensity of their retaliation. The computer then delivered back to group A a sound that was exactly equal to the size of the retaliatory stimulus: a situation characterised in Biblical terms by 'an eye-for-an eye'. The smart reaction would be only to retaliate with a very mild sound, and sober participants did just this. Such optimal behaviour involved the participants assessing the consequences of their actions and using this to control their reaction. Intoxicated participants administered retaliatory sounds that were almost *three times* the intensity of those administered by the sober participants – they did not show the advantageous restraint of sober participants.

In a quite different context, someone might be excessively generous under the influence of alcohol, discounting the negative consequence of being out-of-pocket in the longer term. Indeed, the size of tip left in a restaurant increases with the amount of alcohol consumed, even when bills of constant size are compared (Lynn, 1988).

4.3   Alcohol and the brain

4.3.1 How neurons work

After drinking, ethanol quickly reaches the cells of the brain and affects them, and thereby alters mental states, mood and behaviour. To understand how this happens, it is necessary to look at some of the neurons of the brain (Figure 4.6a overleaf) and how they communicate with each other. Neurons convey signals by means of electrical pulses called **action potentials** (Figure 4.6b–e). When a neuron is active it produces action potentials at a certain frequency or rate.

◆ What does Figure 4.6 show to be the change in electrical events as the neuron becomes *more* excited?

◆ There is an *increased frequency* of producing action potentials.

(When the frequency of action potentials shown by a neuron decreases, it is said to be 'less excited'). Figure 4.7 shows neurons conveying signals from one to another (in the form of action potentials) from the eyes to the brain, and thence to trigger muscles. This section will focus on events within the brain.

A series of interconnecting neurons is shown in Figure 4.8, together with the **synapses** (sigh-nap-sez) that form the point of communication between them. The part of a neuron that looks like a wire is termed the **axon**. It is along axons that action potentials travel, thereby conveying signals over distances within the brain and between the brain and the rest of the body. When an action potential arrives at a synapse, it causes the release of a chemical **neurotransmitter**. This is the means of communication between neurons. The neuron that stores and releases neurotransmitter at a synapse is termed the **presynaptic neuron**. The neuron that has **receptors** for the neurotransmitter is termed the **postsynaptic neuron**. The components that make up a synapse are:

- part of the presynaptic neuron, which stores chemical neurotransmitter,

- part of the postsynaptic neuron, which has receptors for this neurotransmitter and

- the synaptic cleft ('gap') between these neurons.

◆

Dendrites are short extensions from the cell body. They are another site of communication between neurons.

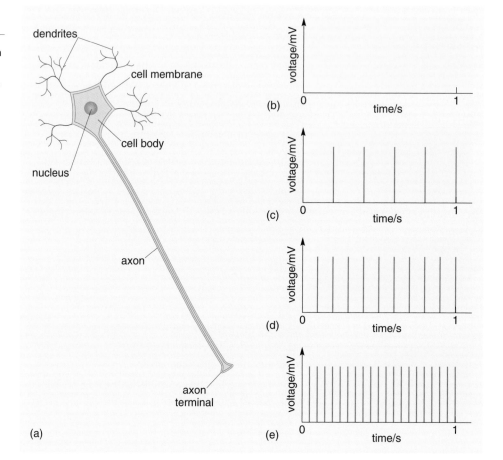

Figure 4.6 (a) Schematic diagram of a neuron; (b)–(e) represent *increasing* levels of excitation as more action potentials are produced. (Source: BSCS)

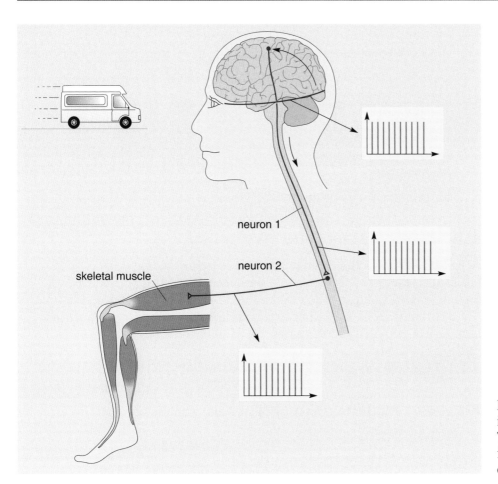

Figure 4.7 The brain showing incoming and outgoing signals. The small graphs show the frequency of action potentials at each location.

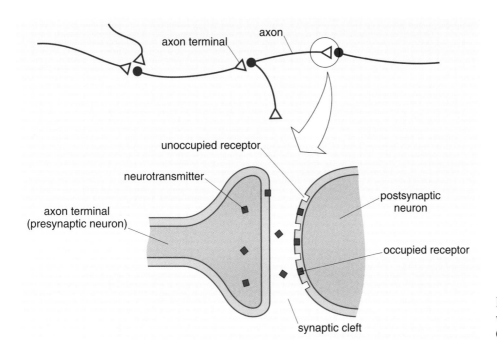

Figure 4.8 Several neurons with synapses between them. One synapse is shown enlarged.

The sequence of actions that take place at a synapse is as follows:

1   An action potential arrives at the end of an axon, termed an **axon terminal**,

2   neurotransmitter is released from the axon terminal of the presynaptic neuron,

3   neurotransmitter rapidly moves across the synaptic cleft,

4   neurotransmitter occupies receptors at the membrane of the postsynaptic neuron,

5   by occupying receptors, the postsynaptic neuron is either excited or inhibited, and

6   the neurotransmitter is inactivated.

Neurons can be characterised by the *type* of neurotransmitter that they store and release when an action potential arrives at the synapse.

At an **excitatory synapse** neurotransmitters *increase* the frequency of action potentials in the postsynaptic neuron. But suppose an action potential arrives at an **inhibitory synapse**, where the neuron is already being excited. The neurotransmitter at the inhibitory synapse *decreases* the frequency of action potentials in the postsynaptic neuron.

4.3.2 The cellular sites of action of ethanol

Of the neurotransmitters important in the alcohol story, glutamate is excitatory (Figure 4.9a), but GABA (gab-ah) (Figure 4.9b) is inhibitory. Part (iii) of

GABA is the scientific short-hand term for gamma-aminobutyric acid.

Figure 4.9   (a) The excitatory action of glutamate: (i) both neurons inactive; (ii) activity in the presynaptic neuron releases glutamate which excites the postsynaptic neuron; (iii) same activity of presynaptic neuron as in (ii) but in the presence of ethanol.

(b) The inhibitory action of GABA at a neuron where an excitatory synapse is also located (activity at the excitatory synapse is constant throughout): (i) inhibitory neuron, which releases GABA is inactive; (ii) inhibitory neuron is now active; (iii) same activity of inhibitory neuron as in (ii), but now in the presence of ethanol.

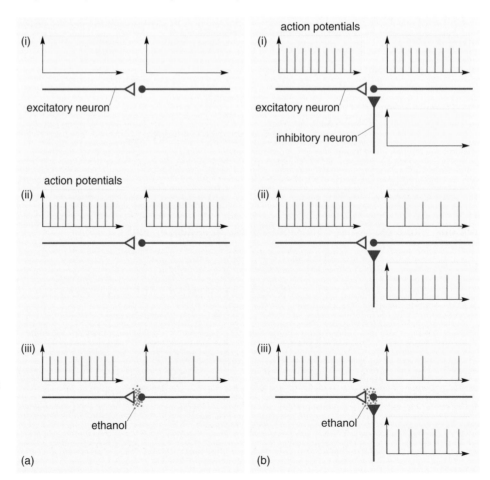

Figures 4.9a and 4.9b show the action of ethanol at these types of synapse.

Some substances have the effect of mimicking the action of a natural neurotransmitter and they are termed **agonists** to the neurotransmitter. Agonists can be administered as medicine in order to give a boost to the activity of particular types of synapse. Conversely, a substance that blocks the action of a natural neurotransmitter is termed an **antagonist** to it.

As shown in Figure 4.10, ethanol attaches itself to certain classes of synapse, at both presynaptic and postsynaptic sites, and *alters the activity at the synapse*:

Ethanol is an *antagonist* to glutamate.

◆ How will this affect the activity of a postsynaptic neuron (Figure 4.9a(iii))?

◆ By antagonising excitatory glutamate, ethanol *decreases* the action potential frequency in this case.

By attaching to sites at a presynaptic neuron that employs GABA, *ethanol triggers the release of GABA*, and by attaching to sites on the postsynaptic neuron, ethanol boosts the actions of GABA. As GABA is an inhibitory neurotransmitter, the frequency of action potentials *decreases even more* (Figure 4.9b(iii)).

◆ What term would relate ethanol and GABA?

◆ Ethanol is an agonist to GABA.

By changing the activity at synapses, ethanol changes the activity patterns of neurons in the brain. This is the basis of the changes in mood and behaviour that are triggered by drinking alcohol.

Ethanol has stronger effects in some parts of the brain than others. In neurons of the type shown in Figure 4.9, activity is lowered by ethanol in both cases for different reasons. This is reflected in an overall lowering of activity in certain brain regions where the synapses are located at a high density and these effects can be better understood in terms of what the different brain regions do. So let us turn now to consider certain locations in the brain.

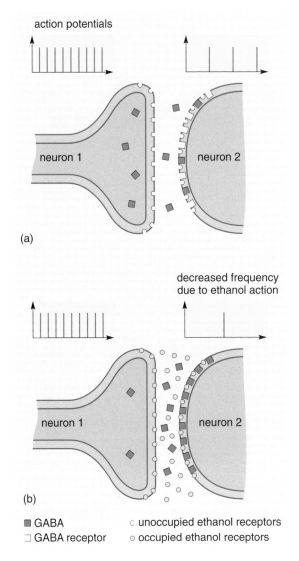

Figure 4.10 Synapse that employs GABA: (a) without ethanol and (b) with ethanol. (Note: neuron 2 is spontaneously active when neuron 1 is not inhibiting it.)

4.3.3 Sites in the brain

Cerebellum

The brain region termed the *cerebellum* (see Figure 4.11) has a vital role in sensory-motor coordination and its information processing is affected by ethanol. Hence, when under the influence of alcohol, motor coordination is disrupted, e.g. when driving a car or trying to walk in a straight line.

However, we will not discuss this further. Instead, this section will focus upon two of ethanol's sites of action, one in the **cortex**, the outer layer of the brain, and

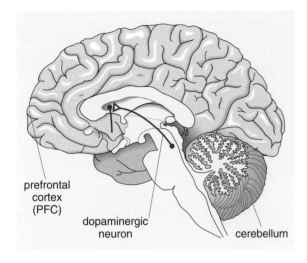

Figure 4.11 Some brain regions of particular interest in studying the effects of alcohol. A dopaminergic neuron is shown.

The neurons employing dopamine are known as 'dopaminergic' (doh-pa-meen-er-jik).

the other, a group of neurons that employ the neurotransmitter dopamine (doh-pa-meen), deeper in the brain (see Figure 4.11).

The group of dopaminergic neurons

Stimuli associated with natural rewards, such as the sight of food when hungry, tend to excite a group of neurons in the brain that use dopamine as their neurotransmitter. When these neurons are excited, the person will tend to seek the reward, for example taking the food, and eating it. By its chemical action on the dopaminergic neurons in this pathway, ethanol *increases* their activity, leading people to seek rewards more actively.

As a result of classical conditioning, stimuli that have been paired with alcohol in the past, e.g. the sight of a glass of wine or the sight of another drinker, acquire the capacity to trigger activity in these neurons and release dopamine. Thereby, alcohol itself can come to form a powerful reward to be sought.

A cortical site

A region of cortex at the front of the brain, termed the **prefrontal cortex (PFC)**, is shown in Figure 4.11. The activity of neurons there is associated with voluntary control of behaviour (self-control) and restraint. Biological evidence suggests that mild to moderate doses of alcohol selectively lower the activity of neurons in the PFC, relative to other regions. This is because of ethanol's actions at synapses located there, which lowers activity particularly in this region.

People who are intoxicated with alcohol are less restrained. Normally, people can anticipate the consequences of their actions and use this as a guide in controlling behaviour. Following intoxication with alcohol (or damage to the prefrontal cortex), there is less ability to anticipate the consequences of behaviour and act accordingly: the person often knows what offensive behaviour is, and yet still goes ahead to act offensively, only to regret it (think back to the students' accounts in Activity 1.3).

Linking back to dopaminergic neurons:

- Alcohol activates dopaminergic neurons, leading people to seek natural rewards.

- Stimuli associated with alcohol also activate the dopaminergic neurons, triggering further seeking and drinking of alcohol.

- Alcohol reduces the activity of neurons in the prefrontal cortex that would normally exert an inhibitory effect on behaviour.

Activity 4.1a uses computer animations to illustrate the action of synapses, and 4.1b shows experimental work in progress on the effects of drinking alcohol.

Activity 4.1a   The nervous system and ethanol

Activity 4.1b   An experiment involving alcohol

Allow about 45 minutes

Now would be the ideal time to study the activities entitled 'The nervous system and ethanol' and 'An experiment involving alcohol' on the DVD associated with this book. The first activity will consolidate your understanding of the first three sections of this chapter. It is an interactive animation showing what happens when ethanol interacts with certain synapses. The effects of alcohol on motor coordination are described and its effects on social inhibitions are related to inhibitory regions of the brain.

The second activity looks at the work of researchers based in the psychology department in the University of Sussex, into alcohol. It highlights some of the important factors that need to be considered when performing experiments with ethanol. On looking at the first part of this sequence, consider the care that needs to be brought to designing a study involving ethanol, e.g. disguising the ethanol and employing a placebo condition. Note the ethical and legal considerations involved in such experimentation. In the second part, Professor Duka describes the role of classical conditioning in triggering alcohol seeking and drinking.

As you watch the sequence, try to answer the following questions:

1   Why is the ethanol disguised by another taste?

2   Why is blood-ethanol level alone not always a sufficiently reliable guide to the expected psychological effects of alcohol?

3   What example of 'alcohol myopia' is revealed in the experiments described?

4   Did any participant show evidence of a placebo effect?

Comments

1   The disguise is so that a participant will not know into which group he or she has been allocated. This knowledge might create an expectation as to what the effect might be, which could influence the actual outcome.

2   The effect can depend upon whether the ethanol level is rising or falling, a so-called 'biphasic effect'.

3   Alcohol tends to bias decision-making in favour of short-term gain and away from decisions that bring long-term benefits.

4   Yes – one participant reported feeling the effects of alcohol even though he had been given only the placebo treatment.

4.4   Associations between alcohol, smoking and sex

There are a range of associations with alcohol and this section considers two of these.

4.4.1 Cigarette smoking

Comparing different individuals, a positive correlation exists between the amount of alcohol that they consume and the number of cigarettes that they smoke (Tiffany, 1990). It could be that smoking triggers drinking alcohol or vice versa (or a third factor might cause both).

The evidence points strongly to there being causal connections between drinking and smoking. Each behaviour tends to trigger activity in the dopaminergic

Figure 4.12 Smoking can trigger the desire to have a drink and vice versa. (Photo: Carol Midgley)

◆

Some of the health effects of smoking are discussed in another book in this series (Midgley, 2008).

neurons and thereby excites the other behaviour (Figure 4.12). Amongst smokers, drinking alcohol increases the motivation to smoke, the craving for a cigarette, and the pleasure derived from smoking (Field et al., 2005). Alcohol also increases the power of cues such as the sight of a packet of cigarettes to attract the drinker's attention. In one study, smokers were asked to judge the pleasantness of pictures, some smoking related (e.g. a person with a cigarette in hand) and others non-smoking related (e.g. someone holding a pen). Alcohol increased the pleasantness rating of the cigarette-related pictures, relative to the non-smoking related pictures. By lifting inhibition, alcohol increases the danger of relapse amongst people who are trying to give up smoking.

As you saw in Section 4.3.3, the effect of ethanol on dopamine is due both to the chemical action itself on neurons in the brain and to classical conditioning (e.g. the sight of the glass of wine having an effect on the neurons). The time at which alcohol has been drunk and a cigarette smoked might be associated in the mind of the smoker, and any history of associations will tend to be self-reinforcing.

4.4.2 Sexual behaviour

Drinking alcohol is a common factor in both voluntary sexual behaviour and sexual assault. Ethanol has a number of effects in the brain that reinforce each other to increase the tendency to engage in sexual behaviour.

◇ Can you suggest two of these?

◆ Ethanol excites the dopaminergic neurons, which cause a person to seek rewards (in this case sex). Ethanol can also reduce the restraining activity of the prefrontal cortex.

Alcohol myopia makes a person less concerned about future events (Cooper, 2006), so if sexual activity is the immediate concern, there may be less likelihood that, for example, pregnancy or AIDS warnings will be thought about (see Rachael's behaviour in Vignette 1.3 and the students in Activity 1.3). In addition, there is some evidence that if a person is under the influence of alcohol, potential partners tend to look more attractive (Jones et al., 2003, and Activity 1.3).

Another massive problem with alcohol is addiction to it, discussed next.

4.5 Addiction to alcohol

4.5.1 What is addiction?

The term 'addiction' has been traditionally used for addictions to chemicals such as heroin, nicotine and so on. Amongst some psychologists, the definition now tends to be cast broader to include such 'out-of-control' activities as excessive gambling, internet use and shopping. By whatever criterion, it is clear that alcohol can be addictive (Vignette 4.1).

Vignette 4.1 Rachael's increasingly heavy drinking and the development of addiction

For many years, Rachael was able to drink heavily and maintain most of the routines of domestic and work-related life. However, she became 'tolerant' to alcohol and needed about three-quarters of a bottle of spirits each day for her to experience the same effects that much smaller amounts used to have. Within the culture of the office she was seen as a '*good sport*' and on social occasions could be relied on to be the '*life and soul*' of the party. From time to time, after a particularly heavy bout of drinking, she suffered from really bad hangovers and had to rearrange her diary in order to sober up. If she didn't have an alcoholic drink for a few days, so-called 'withdrawal symptoms' made her feel even worse.

In some ways the alcohol was useful to her; it helped her to cope in tense and potentially difficult situations. In this way it acted as a reinforcer, so that drinking alcohol became associated in Rachael's mind with relaxation, coping and achievement. However, the reality was rather different and the alcohol tended to blunt her cognitive ability and in particular her own self-awareness.

So, was Rachael addicted to alcohol? People working in medicine and other caring professions have attempted to analyse when someone might be labelled as having an **addiction** or chemical dependence. Look at the description in Table 4.1 which attempts to establish whether a person is addicted to any chemical (including alcohol). If the person has three or more of the behaviour patterns from Table 4.1 over a period of at least a year they may be judged to meet the official criteria for addiction. It certainly seems as though Rachael had several of these patterns in her life. Although at one level Rachael seemed to be holding her life together, she became reliant on others (her husband and her secretary), to support her and to collude in her increasingly destructive habit.

Table 4.1 Official definition of addiction (DSM-IV). (Source: American Psychiatric Association, 1994)

DSM-IV stands for the *Diagnostic and Statistical Manual of Mental Disorders* (4th edition), published in 1994.

1 *Tolerance*: need to drink/use more to get same effect, or diminished effect with same amount.

2 *Withdrawal*: physical/emotional withdrawal symptoms, or drinking/using more to relieve or avoid withdrawal symptoms.

3 *Loss of control*: drinking/using more, or for longer, than intended.

4 *Attempts to control*: persistent desire or efforts to cut down or control drinking/ use of the substance, including making rules for self about when, where, what to drink/use, etc.

5 *Time spent on use*: spending a great deal of time getting the substance, drinking/using it, or recovering from drinking/use.

6 *Sacrifices made for use*: giving up or reducing social, work, or recreational activities that are important to the person because of conflicts with drinking/using.

7 *Use despite known suffering*: continuing to drink/use despite knowing one has a physical or psychological problem that is caused or made worse by drinking/using.

4.5.2 Characteristics of addiction

A fundamental characteristic of addiction is the person's experience of a *compulsion* to engage in an activity (Lyvers, 2000). Expressed in other words, behaviour has got 'out of control', and can disrupt work and family life. Addicts often mean to show restraint, but despite this the addiction wins the struggle. In alcoholics, drinking alcohol tends to create the desire for more alcohol.

Depression or irritability can be experienced when the person is denied access to the activity. Addiction is associated with **craving**, i.e. intense conscious occupation with thoughts of the object of the addiction. This is particularly the case, when (a) the addictive activity is thwarted or (b) temptation is being actively resisted (Tiffany, 1990).

Addiction is associated with the notion of **dependence**: addicts come to *depend upon* the drug for their 'normal' mental functioning. For Rachael, addiction is also associated with **tolerance**: over time, a need for an increasing amount of alcohol to obtain the same level of intoxication.

According to the drug and circumstances of taking it, when the drug is not taken there can be **withdrawal symptoms**. These consist of characteristic physiological signs associated with a negative mood. In the case of alcohol, withdrawal symptoms include increased heart rate and body temperature, and shaking, termed a 'tremor'. There is lack of sleep (insomnia) as well as anxiety. Although addicts may seek their drug in order to avoid withdrawal symptoms, this does not explain all of the motivation for seeking drugs (Robinson and Berridge, 1993); for example, the most intense cravings for a drug often do not correspond to when withdrawal symptoms are experienced. Withdrawal symptoms cannot explain how the addiction arose in the first place (Lyvers, 2000).

Even if the person recovers from addiction, there is a high risk that they will revert to using the drug later. Stress commonly triggers relapse.

4.5.3 Explanations of addiction

There are at least two psychological explanations (or 'orientations') for addiction. These have sometimes been seen as rival explanations but there are elements of truth in each. According to the *exposure orientation*, addiction arises simply as a consequence of frequent use of a drug. This might fit the observation that exposure to alcohol (or nicotine or other drugs) during adolescence is associated with a relatively high risk of later alcoholism (Hingson et al., 2003). Vignette 1.1 on Rachael's early drinking experiences exemplifies this.

According to the *adaptive orientation*, addicts try to use a drug to solve an emotional problem in their personal life. The drug is a kind of self-medication to treat psychological distress. Alas, excessive drug intake increases their unhappiness and increasing amounts of drug are needed, forming a vicious cycle.

Psychological investigations have found that depression and anxiety disorders, anti-social tendencies, and social factors of alienation, such as unemployment, are associated with a high risk of addiction.

Anyone who has given up smoking will know how their minds were dominated by thoughts of cigarettes.

4.5.4 Treatment

There are a number of ways of trying to treat addiction to alcohol, some of them directly alcohol-related and some more indirect. A combination of treatments can be tried.

Addicted people often have an underlying psychological problem that they are trying to self-medicate, such as depression, chronic stress or lack of self-esteem. In Section 1.3 we noted that alcohol intake is increasing in developing countries that are subject to stress associated with life in the 'human zoo'. Stress is also a major contributor to relapse (Weiss, 2005). Psychological interventions, such as counselling, can target such problems. Self-help groups try to create the conditions of mutual support and fellowship that help the addicted person to change.

Some drug treatments try to target the neurotransmitters in the brain that underlie alcohol seeking (Buonopane and Petrakis, 2005) and thus block the rewarding effects of alcohol. This can be highly problematic in use, since any neurotransmitter involved in seeking rewards (e.g. alcohol) will also be involved in a variety of other roles, and undesirable side-effects, such as sleepiness, can result. The neurotransmitter serotonin (seer-oh-toe-nin) is a potential target for drug treatment in the case of both depression and addiction.

◈ What class of chemical intervention blocks the effect of a neurotransmitter?

◆ That described by the term 'antagonist' (Section 4.3.1).

The evidence points to neurons that employ neurotransmitters of the class termed *opioids* as forming the biological basis of reward and pleasure (Berridge, 2003). Therefore, if an intervention were able to disrupt synapses that employ opioids, maybe the attraction of alcohol could be reduced. A drug of this class is naltrexone, which antagonises opioid neurotransmission.

The drug termed acamprosate appears to target glutamate and GABA, and is showing some success in treating alcoholism (Soyka and Roesner, 2006).

Another class of intervention is to try to create unpleasant consequences of drinking. The logic here is that the expectation that alcohol brings pleasure will alter to the expectation that alcohol brings nausea, and it will lose its attraction.

◈ You met an example of such a drug in Section 2.4.1. Can you recall its name and how it acts?

◆ The drug is Antabuse, which acts by preventing the conversion of acetaldehyde into acetic acid.

However, getting people who are addicted to alcohol to take medicines regularly is difficult as it takes persistence and will-power for a person to voluntarily undermine the crutch that is 'supporting' them.

Now try Activity 4.2.

Activity 4.2   Alcoholics Anonymous

Allow 30 minutes

Now would be the ideal time to study the video entitled 'Alcoholics Anonymous' on the DVD associated with this book.

This video describes the ethos underlying Alcoholics Anonymous (AA), and follows the case studies of three very different members. As you watch, consider the issues of whether alcohol can serve as a kind of crutch to support a person in psychological pain. Consider also the evidence that helping others and gaining a meaning in life can serve to undermine the need for such a chemical crutch.

Summary of Chapter 4

4.1   Both psychological and biological sciences have a role in explaining why people drink alcohol and its effects, captured by the term 'psychobiological' perspective.

4.2   A complex of internal and external factors determines the tendency to drink alcohol or to decline it. These include anxiety and frustration, the presence of other drinkers, conditional stimuli and mental projections to the future and past.

4.3   The term 'alcohol myopia' refers to the tendency of alcohol to cause a restriction of the focus of attention to the more immediate events.

4.4   At a cellular level, ethanol exerts an effect at synapses, e.g. those that employ the neurotransmitters glutamate and GABA.

4.5   In terms of brain regions, ethanol tends to increase activity in dopaminergic neurons and lower activity in the prefrontal cortex.

4.6   By its intrinsic chemical effects and by conditioning, alcohol encourages smoking.

4.7   Alcohol can increase tendencies to violence and to engage in risky sexual behaviour.

4.8   Addiction to alcohol is associated with alcohol tolerance, compulsion, craving and withdrawal symptoms on halting intake; the effects on the lives of people who are addicted can include loss of relationships, jobs, homes and self-esteem.

4.9   Chemical treatments for alcohol addiction include the use of antagonists to particular neurotransmitters and Antabuse; peer-support groups such as Alcoholics Anonymous may also be effective.

4.10   Research studies involving subjects who drink controlled doses of alcohol must be carefully designed to ensure their safety; the use of a placebo in place of alcohol demonstrates that some of the apparent effects of alcohol are due to the subject's expectations.

Learning outcomes for Chapter 4

After studying this chapter and its associated activities, you should be able to:

LO 4.1   Define and use in context, or recognise definitions and applications of, each of the terms printed in **bold** in the text. (Questions 4.1, 4.2 and 4.3)

LO 4.2   Explain how the attraction of alcohol and the triggering of drinking depend upon a combination of internal and external factors, including classical conditioning. (Question 4.1 and DVD Activities 4.1a and b)

LO 4.3   Describe the effects of alcohol on memory and in terms of alcohol myopia and the lifting of inhibition; illustrate these effects by reference to the experiment in Activity 4.1b and the way in which it was carried out. (Question 4.1 and DVD Activities 1.3 and 4.1b)

LO 4.4   Describe the effects that ethanol has at the level of synapses and thereby explain how it can affect the activity of the nervous system. Relate this to the effects drinking has on mood and behaviour. (Question 4.2)

LO 4.5   Describe what is meant by 'addiction' and how addiction to alcohol might be better understood, and treated. (Questions 4.2, 4.3 and DVD Activity 4.2)

Self-assessment questions for Chapter 4

You have had the opportunity to demonstrate LOs 4.2 and 4.3 by answering questions in DVD Activity 4.1a and after Activity 4.1b.

Question 4.1 (LOs 4.1, 4.2 and 4.3)

What would it mean to claim that a placebo effect is associated with alcohol myopia and the lifting of inhibition?

Question 4.2 (LOs 4.1, 4.4 and 4.5)

Based upon an understanding of the neurotransmitters most closely linked to alcohol seeking and drinking, which of the following would be suggested as a form of treatment for excessive alcohol intake and why? (i) An agonist to dopamine. (ii) An antagonist to dopamine.

Question 4.3 (LOs 4.1 and 4.5)

Was Rachael addicted to alcohol, according to the definition in Table 4.1? Justify your conclusions by drawing on information in any of the 'Rachael' vignettes in this or earlier chapters.

CHAPTER 5

HOW ALCOHOL CAUSES SHORT- AND LONG-TERM HARMFUL EFFECTS

5.1   Introduction

This chapter investigates some of the harmful effects that a high level of blood-ethanol can have on the body: both short-term problems such as 'hangover', and long-term health problems that are associated with regular heavy drinking. Whilst this chapter is primarily about the biological effects that ethanol has on various organs of the body, it is important to remember that the socioeconomic effects of heavy drinking (Chapter 1) are also very serious (Paton, 2005). Figure 5.1 summarises some of the increased risks a drinker faces as the ethanol concentration in their blood increases.

The main acute effects of ethanol are on the nervous system, causing increased confidence and heightened mood (Section 4.2.2) leading to increased risk of accidents, violence and socially inappropriate behaviour (Section 1.5). Alcohol also disrupts the ability of the brain to coordinate between sensory input and motor output which leads to slurred speech, less controlled movement, delayed reaction times and errors of judgement (Section 4.2.3). Tests on drivers have shown that steering problems begin at blood-ethanol concentrations (BAC) of about 20 mg/100 ml, and the risk of being involved in a traffic accident more than doubles at BAC of 80 mg/100 ml, which is the legal limit for driving in the UK (Table 5.1 overleaf). If this concentration doubles to 160 mg/100 ml, the risk of being involved in an accident increases more than 10-fold.

◆ In the experiments mentioned in Section 4.2.2, experienced drivers misjudged gaps with a concentration of 50 mg ethanol in 100 ml of blood (BAC 50 mg/100 ml). From Table 5.1, in which countries would it be legal for drivers to exceed this amount?

◆ Drink-drive limits are set at *more* than 50 mg/100 ml in Cyprus (South), Ireland, Luxembourg, Switzerland and the UK.

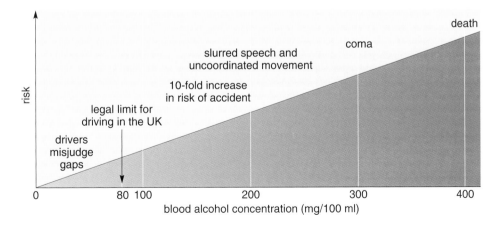

Figure 5.1   Some of the risks associated with increasing ethanol concentrations in the blood. (Source: Paton, 2005)

Table 5.1 European drink-drive blood-alcohol limits (Source: Safe Travel, 2006)

Country	Limit (mg/100 ml)	Country	Limit (mg/100 ml)
Austria	50	Lithuania	0
Belgium	50	Luxembourg	80
Bulgaria	50	Macedonia	50
Croatia	50	Malta	0
Cyprus (North)	50	Netherlands	50
Cyprus (South)	90	Norway	20
Czech Republic	0	Poland	20
Denmark	50	Portugal	50
Estonia	0	Romania	0
Finland	50	Serbia and Montenegro	50
France	50	Slovakia	0
Germany	50	Slovenia	50
Greece	50	Spain	50
Hungary	0	Sweden	20
Ireland	80	Switzerland	80
Italy	50	Turkey	50
Latvia	50	UK	80

If BAC levels increase to around 200 mg/100 ml then speech becomes slurred and coordination of movement is impaired. Coma (loss of consciousness) is likely to occur above 300 mg/100 ml, and concentrations above 400 mg/100 ml are likely to be fatal as a result of disruption of heart and lung function, or inhalation of vomit (recall Section 1.5.1).

There are variations in the way that people respond to alcohol and some of these are described in Box 5.1.

Box 5.1 (Explanation) Variations in alcohol tolerance

In some individuals, the main acute effects of drinking an alcoholic drink are rather different from those just described – they rapidly become quite unwell and experience what is known as a 'flush' reaction. The symptoms of this reaction are redness of the face and chest, rapid heart rate, dizziness, nausea, nasal congestion and pulsating headaches. More severe reactions can cause breathing problems and low blood pressure. Individuals who experience these effects accumulate a higher level of acetaldehyde in their blood than those who do not experience these effects.

You saw in Section 2.4.1, that the enzyme alcohol dehydrogenase (ADH) which converts ethanol into acetaldehyde and the enzyme aldehyde dehydrogenase (ALDH) which converts acetaldehyde into acetic acid, can each exist in a number of different isoforms which work at different rates. The particular combination of isoforms that individuals have, varies according to the population to which they belong. The production of all proteins in the body, including enzymes, is controlled by *genes*, individual units of the inherited genetic material, DNA. Genes are inherited from the parents, and everyone inherits a complete set of genes, including those that direct the production of ADH and ALDH. However, people from different populations have slightly different variants of each gene. This is why certain ethnic groups have a relatively high proportion of people who become quite ill after drinking small amounts of alcohol – because of increased levels of acetaldehyde in their blood. More than 75% of Japanese people who drink report flushing, compared with 5–10% of Caucasians (Arnon et al., 1995).

◆ Genes and DNA are discussed in another book in this series, (Parvin, 2008)

◆ If unusually high levels of acetaldehyde accumulate following an alcoholic drink, how could this have been caused by (a) the rate of reaction of an ADH isoform, and (b) the rate of reaction of an ALDH isoform?

◆ ADH promotes acetaldehyde production, so an ADH isoform with a *high* rate of reaction would cause more acetaldehyde to accumulate than one with a slower rate. An ALDH isoform with a *slow* rate of reaction removes acetaldehyde slowly, allowing high levels of acetaldehyde to accumulate.

As noted in Section 2.4.1, the overall rate of ethanol metabolism depends on many factors and ranges from a decline in BAC of less than 10 mg/100 ml per hour to over 40 mg/100 ml per hour for different people.

5.2 Hangovers

'Beer is the reason I get up every afternoon.'

In certain cultures, an evening of heavy drinking is a regular social activity and the ill-effects suffered the following morning are accepted as an inevitable part of life. In Chapter 1 the economic cost of *alcohol-related absence* was mentioned. This is frequently caused by workers experiencing symptoms of 'hangover' which is the term used to describe the collection of symptoms that occur in drinkers on the day following a heavy drinking session, once the ethanol has been cleared from the blood (Figure 5.2). Even when a hangover is not severe enough to cause absence, it may severely impair the ability of a person to function effectively in the workplace. The economic impact of alcohol-related illness is dominated by these short-term productivity deficits, with chronic alcohol-related diseases only accounting for a small proportion of this cost.

Figure 5.2   Source: Cartoonstock.com.

The list below shows the most reported symptoms of hangover, with the commonest first:

- headache
- poor sense of well-being
- diarrhoea
- anorexia (lack of appetite)
- tremor (trembling hands)
- fatigue
- nausea

Symptoms vary enormously between people and episodes, making research into this condition difficult, even without considering the ethical issues of deliberately making people ill. A variety of physiological mechanisms have been proposed that could reasonably explain the occurrence of hangovers, but without much evidence to back them up. We will consider each of the possibilities briefly.

5.2.1 Physical disturbances

Dehydration

◆ How can drinking alcohol result in dehydration?

◆ Ethanol inhibits the release of vasopressin; this results in an increase in the volume of urine produced by the kidneys and can lead to dehydration (Section 3.4).

Although vasopressin levels return to normal during hangover, once ethanol is no longer present in the blood, additional fluid loss may occur due to sweating, vomiting and diarrhoea. Symptoms of dehydration include many of those associated with hangovers (dizziness, light-headedness, weakness, thirst, etc.).

Gastrointestinal disturbances

At a high concentration, ethanol damages cells on contact (Box 5.2) and so can cause irritation of the stomach and intestinal lining resulting in inflammation.

Ethanol also increases the production of gastric acid and intestinal secretions, and the after-effects of these processes include abdominal pain, nausea and vomiting, all of which can be associated with hangovers.

> **Box 5.2** (Enrichment) The antibacterial action of ethanol
>
> Ethanol can act as a disinfectant because it damages cells it is in contact with. It is an effective disinfectant and is used at 70% vol. strength in hospital swabs and laboratories to kill bacteria.
>
> In previous centuries, weak beers and ciders were often brewed by, or given as a ration to, the peasant workforce, and were the drink of choice. Because of the antibacterial action of the ethanol, they were safer to drink than the often-contaminated water.

Low blood sugar

The level of glucose which is present in the blood is carefully maintained in order to ensure that cells receive a continuous supply of fuel for metabolic processes. Excess glucose from the diet is stored in the liver and released in a carefully regulated manner to maintain the level in the circulation. Ethanol consumed in large amounts disrupts this balance and this can result in low blood sugar levels a few hours later. Since glucose is the primary source of energy for cells in the brain, this could explain hangover symptoms such as fatigue, weakness, and mood disturbances.

5.2.2 Psychological and sleep disturbances

The severity of hangover symptoms has also been associated with particular personality traits. For example, some research has indicated that individuals with personality traits that predispose them to a risk of alcoholism, experience more severe hangover symptoms than other people.

Although alcohol acts as a sedative, the sleep it induces can be of poorer quality and shorter duration than normal. Ethanol interferes with the action of key neurotransmitters, in particular GABA and glutamate, to cause an overall sedative effect. The body counterbalances this by changing the number of GABA and glutamate receptors present at the synapses, increasing the number of glutamate receptors and decreasing the number of GABA receptors. Ethanol is removed relatively quickly from the body, but it takes longer for the number of receptors to return to normal, leaving the nervous system in a temporary unbalanced over-excited state.

◆ What are the physiological signs associated with alcohol withdrawal?

◆ Elevations in heart rate and body temperature, tremor, insomnia and anxiety (Section 4.5.2).

Because the symptoms of hangover share many of the characteristics of alcohol withdrawal symptoms, a possible interpretation is that hangover may be a mild form of alcohol withdrawal.

5.2.3 Chemical factors

Ethanol metabolites

The suggestion that acetaldehyde accumulation is involved in hangovers is largely due to the observation that high concentrations of acetaldehyde in the blood give rise to toxic effects which resemble hangovers (rapid pulse, sweating, nausea, etc.).

Metabolites are the chemicals produced during 'metabolic' processes, so acetaldehyde and acetic acid are ethanol metabolites.

◆ Why would acetaldehyde accumulate in the body after heavy drinking?

◆ Ethanol is converted into acetaldehyde by the enzyme ADH, and then ALDH catalyses the breakdown of acetaldehyde into acetic acid (Section 2.4.1). If the first reaction occurs at a faster rate than the second, then acetaldehyde accumulates.

However, high concentrations of acetaldehyde in the blood are *not* detected during hangovers in most people, so there is no direct evidence that acetaldehyde is the root cause.

Other chemicals

Studies have shown that purer drinks (such as vodka and gin) cause fewer hangover symptoms than drinks rich in *congeners* (whisky, brandy, red wine). Congeners are the chemicals that provide characteristic taste, aroma and colour. However, pure ethanol can cause hangovers, so congeners can only be a contributing factor. One chemical that has been identified as potentially important in hangover is methanol.

Methanol is a similar molecule to ethanol but contains one fewer carbon atom.

$$H-\underset{\underset{H}{|}}{\overset{\overset{H}{|}}{C}}-\underset{\underset{H}{|}}{\overset{\overset{H}{|}}{C}}-O-H \qquad H-\underset{\underset{H}{|}}{\overset{\overset{H}{|}}{C}}-O-H$$

ethanol   methanol

The internationally accepted chemical names for formaldehyde and formic acid are methanal and methanoic acid, respectively. They are not always used and here the traditional, more familiar names are used.

Methanol is absorbed and metabolised by the same mechanisms as ethanol, but ethanol is *preferentially* metabolised when both substances are present. Methanol levels in the blood therefore remain high after ethanol levels decrease, possibly explaining the delayed onset of hangover (Jones, 1987). The products of methanol breakdown are *formaldehyde* and *formic acid*, both of which are very toxic: high concentrations can cause blindness and death. Drinks which are associated with increased hangover symptoms contain high levels of methanol (e.g. brandy, whisky).

$$H-C \overset{O}{\underset{H}{\lesseqgtr}} \qquad H-C \overset{O}{\underset{O-H}{\lesseqgtr}}$$

formaldehyde   formic acid

Methanol poisoning is treated by administering large doses of ethanol; the methanol is thus metabolised more slowly, preventing the build-up of toxic formaldehyde and formic acid. This approach might form the basis for theories that consuming more alcohol is a way to overcome a hangover.

Hangover treatments

There are numerous treatments reputed to alleviate the symptoms of hangover, ranging from anecdotal folklore to costly pharmaceutical products. Hangover symptoms abate with time, but vary widely between individuals and occasions, so trials of remedies are of limited reliability. Whilst these remedies may not have been subject to systematic scientific testing, some are reasonable approaches based on what is understood about hangover physiology:

- Consumption of food containing the sugar fructose, such as fruit or fruit juices, or bland food rich in carbohydrates, may help to counter symptoms arising from low blood sugar.

- Sleeping is likely to relieve symptoms associated with fatigue.

- Drinking copious volumes of water reduces dehydration.

- Anti-inflammatory drugs such as aspirin are commonly used to relieve the symptoms of headache, but there is a risk that since they can irritate the gut lining they might compound alcohol-induced stomach disturbances.

- Caffeine is traditionally widely used and may counteract fatigue symptoms.

- More alcohol (Box 5.3 overleaf) may alleviate mild withdrawal symptoms, and/or allow methanol time to clear the system without producing high concentrations of formaldehyde and formic acid. However, any immediate relief would be overshadowed by the re-initiation of the whole process.

A recent study concluded that 'no compelling evidence exists to suggest that any complementary or conventional intervention is effective for treating or preventing the alcohol hangover' and suggested that an effective intervention would remain elusive until the underlying biology was better understood (Pittler et al., 2005). Hangovers are unpleasant experiences that cause disruption to people's lives and can impair their ability to perform tasks such as working or driving effectively. The most effective way of reducing hangover symptoms is to avoid drinking excessive amounts of alcohol, especially drinks containing the highest levels of congeners.

Box 5.3 (Enrichment) 'Hair of the dog'

This phrase is used to describe the consumption of a small alcoholic drink in order to cure hangover. The expression alludes to an old belief that an antidote to having been bitten by a mad dog was to place some burnt hair from the same dog onto the wound. The phrase is also used in Hungary 'kutya harapást szörével' translated as (you may cure) 'the dog's bite with its fur'.

Phrases to describe the concept of drinking to alleviate hangover symptoms are also found elsewhere, such as the French 'rallumer la chaudière' (re-light the boiler) or the Danish 'rejse sig ved det træ, hvor man er faldet' (you should get up next to the tree where you fell).

5.3 Long-term problems from chronic alcoholism

Chronic alcoholism, the excessive and habitual consumption of alcohol, results in many health problems. Three are discussed here – the damage caused to the drinker's liver and to the nervous system, and the effect of alcohol consumption during pregnancy on the fetus. Other health problems such as cardiovascular disease and cancers were summarised in Table 1.7, but not discussed in detail.

5.3.1 Alcoholic liver disease

Rachael's story now illustrates the start of damage to the liver (Vignette 5.1).

Vignette 5.1 Rachael develops health problems as the result of her continued drinking

Rachael continued to drink heavily throughout her time as a manager in the travel company. For many years she was able to cope with the heavy demands of her job without apparently developing any harmful effects. Indeed it seemed as though the alcohol had a positive effect on her employment prospects. The other managers enjoyed her company in the pub after a tense day at the office and she enjoyed quite a reputation as a woman who could drink as much as the men.

When Rachael was 45 years old her company asked her to re-locate to their new headquarters in Milton Keynes. As part of her application for insurance to cover her new mortgage Rachael went to see her own doctor for a medical check. On routine questioning Rachael disclosed that she often had abdominal discomfort, occasionally felt nauseous and had frequent diarrhoea.

During the medical examination Rachael disclosed that she was drinking the equivalent of four units of alcohol each day, usually more at the weekends (in fact it was much more). Her doctor calculated that this represented approximately 32 units each week, far in excess of the UK Government's recommended *maximum* of 14 units per week for women (21 units for men).

On examination her liver was found to be slightly enlarged, but she had no other external signs of liver disease. The doctor took a blood sample that was sent off for analysis.

Because many people tend to underestimate their alcohol consumption when questioned (Section 1.2), some health workers use specific sets of questions when screening for excessive alcohol intake (Walsh and Alexander, 2000). A popular version is called the CAGE questionnaire:

C   Have you ever felt the need to cut down your drinking?

A   Have you ever felt annoyed by criticism of your drinking?

G   Have you ever felt guilty about your drinking?

E   Have you ever taken a drink (eye opener) first thing in the morning?

A positive answer to two or more of these questions suggests an excessive alcohol intake.

After a week Rachael went back to see her doctor for the results of her blood tests. The doctor informed her that two of the commonest signs of liver damage related to excessive alcohol consumption had been detected in her blood sample.

Making a diagnosis

Rachael's liver disease was picked up almost by chance at a fairly early stage. The doctor was able to make a diagnosis of liver malfunctioning from her medical history and symptoms that were discussed, and the examination which showed an enlarged liver. The blood tests confirmed that Rachael's liver wasn't working very well, but this falls short of establishing a firm diagnosis. An accurate diagnosis can only be made by techniques that differentiate between the different reasons why her liver may be working less than optimally. A liver biopsy (Figure 5.3, overleaf) was used to make certain that Rachael's liver disorder was related to her alcohol consumption.

At this stage it was confirmed that her diagnosis is known as *alcoholic fatty liver*.

Rachael was upset by the diagnosis, but relieved to hear from her doctor that this condition could be reversed if she stopped drinking alcohol. She resolved to stop, or at least severely restrict, her alcohol intake.

During a liver biopsy a small hollow needle is inserted through the skin and into the liver (Figure 5.3). A sample of liver tissue is extracted that can be examined under a miscroscope. Biopsies may be used to diagnose a condition, and also to monitor its progress in future years.

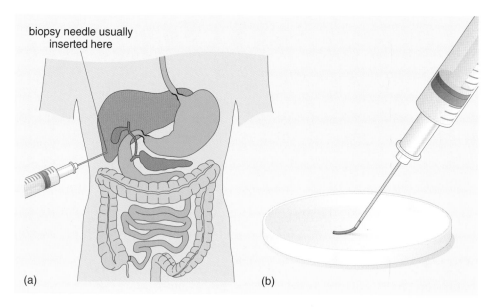

(a)   (b)

Figure 5.3 Taking a liver biopsy. (a) Indicates the position of the liver and the location through which the biopsy needle is usually inserted. (b) A small slender core of tissue is removed.

Perhaps the best known long-term harmful side-effect of drinking excessive alcohol is damage to the liver. **Alcoholic liver disease** is categorised into three progressive stages; fatty liver, hepatitis and cirrhosis. **Fatty liver** is an early and reversible consequence of excessive alcohol consumption during which fat accumulates within the cells of the liver (Figure 5.4). The mechanisms by which this occurs are complex and still under investigation, but they include the release of fats from adipose tissue (cells where fat is normally stored), reduced fat breakdown in the liver, and in cases of chronic alcoholism other nutritional deficiencies play a role.

◆ Why might there be nutritional deficiencies in chronic alcoholism?

◆ Ethanol has a high energy content, so can supply much of the daily energy requirements of a heavy drinker. However, unlike other food sources, it provides none of the other essential nutrients (vitamins, proteins, etc.) so malnutrition is common in alcoholism (Section 2.6).

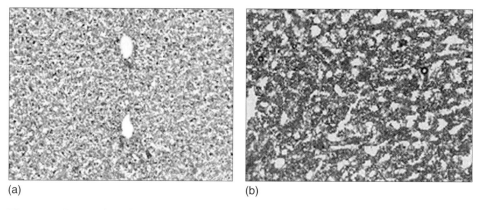

(a)   (b)

Figure 5.4 At the microscopic level it is possible to detect the presence of fat in thin slices of liver tissue using a red dye. (a) The cells in a sample of normal liver contain little fat compared with (b) cells in a sample of fatty liver. Magnification × 100. (Photos: Hilary MacQueen)

Fatty liver itself does not cause long-term damage to the liver and is reversed by abstaining from drinking alcohol. However, it can be an important early indication that harm is being done and that continued excessive alcohol consumption could lead to the more serious conditions of hepatitis and cirrhosis (Figure 5.5):

Hepatitis means 'inflammation of the liver' and can range from being mild (only detectable through blood tests) to severe, causing sickness, jaundice (yellowing of the skin) and pain. Very severe hepatitis can lead to liver failure which is often fatal.

Cirrhosis is a gradual and *irreversible* change in about 10% of chronic heavy-drinkers whereby liver cells are replaced by scar tissue. Not only does this decrease the ability of the liver to perform its many essential biological functions, it also disrupts the blood flow through the liver tissue which causes serious complications such as damage to the spleen (an organ involved in blood maintenance) and the blood vessels of the gut (as blood pressure increases in the portal vein).

Because the liver is vital for a wide range of functions including digestive processes, regulation of blood glucose levels, storage of vitamins and break-down of noxious substances, the disruption caused by cirrhosis affects many body systems. Death from cirrhosis of the liver tends to be due to liver failure or sudden catastrophic bleeding from the disrupted blood vessels of the gut.

Relatively little is understood about the precise changes that occur within the individual cells of the liver that lead to damage, inflammation and cirrhosis. The observation that relatively few individuals who drink excessively suffer serious liver injury (hepatitis and cirrhosis are rare, whereas most heavy drinkers will have fatty liver) suggests that other factors, such as obesity and genetic makeup, may be important in addition to alcohol (Reuben, 2006). In addition, deficits in vitamins and other essential nutrients can lead to disruption of the chemical reactions in the body which maintain healthy tissues and repair damage, so it is possible that malnutrition could contribute to the cause or the progression of alcoholic liver disease.

Figure 5.5 (a) Comparison between samples of liver from three people showing normal liver, fatty liver and cirrhosis. (Photo: Arthur Glauberman/Science Photo Library)

Several potential mechanisms have been proposed that attempt to explain how the presence of ethanol (or its metabolites) could cause damage to liver cells; however, research in this area is difficult due to the enormous number of interlinked chemical reactions that simultaneously occur within cells.

Understanding the molecular processes within liver cells that lead to alcoholic liver disease in some individuals and not others could allow the development of therapies to reduce or perhaps even reverse the harmful effects of drinking alcohol. Currently some alcoholic liver diseases may be reversed with abstinence, but cirrhosis is *irreversible* and treatment (Section 5.6) is focused on slowing progression and reducing complications.

5.4 Nervous-system damage

Chronic consumption of high levels of alcohol can cause irreversible damage to the nervous system. The majority of people with chronic alcoholism have some degree of **dementia**, which is a general loss of intellectual abilities including memory, judgement and abstract thinking, as well as personality changes. The general effect seems to be a shrinkage of brain tissue, as revealed by brain imaging techniques (Figure 5.6) or post-mortem studies, the extent of which correlates with the amount of alcohol consumed (National Institute on Alcohol Abuse and Alcoholism, 2000). In particular shrinkage is extensive in the

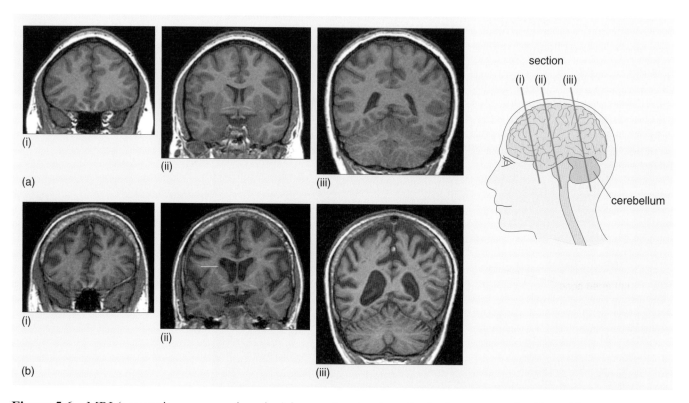

Figure 5.6   MRI (magnetic resonance imaging) image showing how the brain structure is affected in a person with chronic alcoholism. (a) Images taken in three places in a healthy brain – these are 'slices' taken through the front, middle and back of the head (see inset). (b) Shows the shrinkage of the brain in images taken at the same three positions in a person with chronic alcoholism. (Photo: Daniel Hommer/National Institute of Alcohol Abuse and Alcoholism)

prefrontal cortex, PFC, which has responsibility for choice, decision-making and regulation of behaviour. It is also present in deeper brain regions associated with memory, and in the cerebellum which is involved in movement and coordination.

Alcoholism is also associated with damage to the peripheral nerves (i.e. those connecting the central nervous system, CNS, with the rest of the body). This causes symptoms such as sensory disturbances (numbness or pain), motor disturbances (weakness and muscle wasting) and some problems with speech, swallowing, heart rate, pupil function, erectile function, breathing during sleep, etc. The mechanism of nerve damage is not clear and could be associated with a direct toxic effect of ethanol on nerves, or indirectly via nutritional deficiencies.

5.5 Fetal alcohol syndrome

There are a range of disorders associated with maternal alcohol consumption during pregnancy which are collectively known as fetal alcohol spectrum disorders, FASDs. The best characterised is **fetal alcohol syndrome, FAS**. FAS is defined by four criteria, the first of which is excessive maternal alcohol intake during pregnancy, the other three being:

1   a characteristic pattern of minor facial abnormalities and other malformations (in particular of the limbs and heart); the characteristic facial features are illustrated in Figures 5.7 and 5.8 (overleaf);

2   growth retardation;

3   central nervous system abnormalities in the fetus/infant.

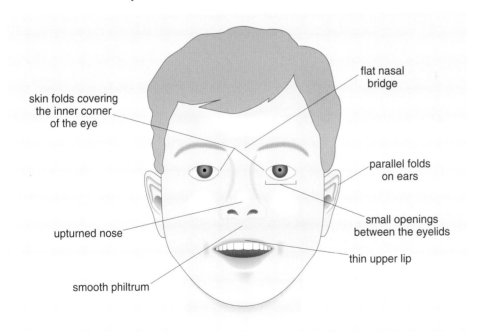

Figure 5.7   Characteristic facial features in a child with fetal alcohol syndrome. Findings may include a smooth philtrum (the vertical groove between the base of the nose and the border of the upper lip), thin upper lip, upturned nose, flat nasal bridge, skin folds covering the inner corner of the eye, small openings between the eyelids and small head circumference. (Source: American Academy of Family Physicians)

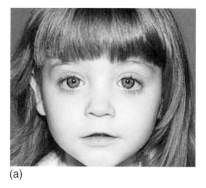

(a)

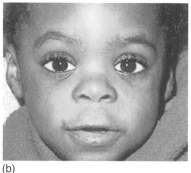

(b)

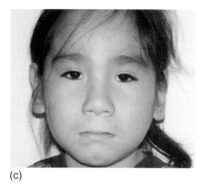

(c)

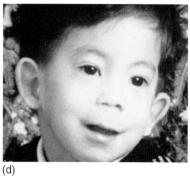

(d)

Figure 5.8 Characteristic facial features in children of different ethnicities with fetal alcohol spectrum disorders illustrating the facial features described in Figure 5.7. (a) Northern European descent. (b) Black. (c) Native American. (d) Biracial (white, black). (Photos: American Academy of Family Physicians)

FAS is considered to be the leading cause of mental retardation in the world and is linked to impaired learning, attention problems, slower reaction times and impaired problem solving, planning and ability to keep track of things (National Institute on Alcohol Abuse and Alcoholism, 2000). Studies from the 1970s and 1980s estimated the incidence rate of FAS in the USA as 1–2 cases per 1000 children born each year. In a population-based study in Seattle in 1997 the rate was nearly 1 per 100 live births. The highest rates to date report 4.6% of 6-year-old children in one community within the Western Cape Province in South Africa (Strömland, 2004).

The mechanisms that result in FAS are complex. Ethanol and acetaldehyde present in the mother's circulation cross the placenta and circulate freely through the blood vessels of the developing fetus. Fetal development involves the division and migration of cells in a carefully orchestrated manner, a process that is disrupted by substances such as ethanol. The developing nervous system is particularly sensitive to disruption.

Not all children who are exposed to alcohol during fetal life are born with FASDs, and there may be several risk factors, involving genetic as well as environmental factors. There are also likely to be many other contributing factors such as nutrition, parental care, drug abuse and smoking which make research into FAS complicated. There are no treatments to reverse the damaging effects of alcohol on the developing brain, although the characteristic facial features lessen during adolescence. A safe threshold of alcohol consumption or critical period during pregnancy when alcohol must be avoided has not been identified. Women who are pregnant, or even those who are just likely to become pregnant, are advised to *abstain* from drinking alcohol. Whilst FAS is entirely preventable by avoiding alcohol during pregnancy, in reality a significant number of women of childbearing age in all western cultures are dependent on alcohol to a certain extent. Treating alcohol dependence in these women is therefore a health care priority for the protection of any future children, as well as for their own health.

5.6 Treatment

5.6.1 What works in helping people to stop or reduce their alcohol intake?

There is at the moment no known cure for alcoholism (see Section 4.5.4 and Activity 4.2). Both psychological and drug treatments are used (Section 4.5.5) in treating excessive alcohol intake (Buonopane and Petrakis, 2005). The plethora of different types of treatment in itself may indicate that there are no guaranteed results, and the condition remains hard to resolve.

Throughout England a mapping exercise (Alcohol Concern, 2002), estimated that there were over 300 advice and counselling services, 100 day programmes and nearly 200 residential programmes specifically established to treat alcohol addiction. The range of treatments available at these centres is very diverse and can be baffling to people seeking help. Many of the non-drug-related forms of support have not been subjected to clinical trials that would be able to demonstrate whether or not they are effective. This is not to say that the therapies and supportive interventions do not work, but just that it is difficult to demonstrate their effectiveness in an objective 'scientific' way.

A proportion of people with alcohol dependency are able to stop using alcohol without the assistance of outside agencies or formal types of help (Klingemann and Sobell, 2001). For Rachael the shock of the realisation that she was harming her own health might be sufficient for her to resolve to make this major change to her behaviour (Vignette 5.2). Research studies that have compared groups of problem drinkers with those in treatment have found that amongst the people not receiving recognised forms of treatment a proportion modify their own drinking habits. A major study in the USA (Weisner et al., 2003), found that 12% of the problem drinkers followed up as a 'control group' and who had received no formal counselling or psychological help, reported that they were still abstinent of alcohol after 12 months, but this is much less than the 57% under formal treatment who were not using alcohol after 12 months.

If stopping drinking alcohol is difficult in the short term, it becomes even more difficult for people to remain abstinent in the longer term. Many people with dependency on alcohol seem to go through several cycles that include periods of controlled drinking or abstinence, followed by problem drinking that leads to further treatment episodes and so on through their lives. Research in Germany (Mann et al., 2005) followed 96 problem drinkers over 16 years, and confirmed

Vignette 5.2 Rachael needs treatment

When she was first given the diagnosis of 'alcoholic fatty liver', Rachael was determined to find out as much as she could about this condition. After several sessions on the internet and following a further consultation with her own doctor she was able to find out the following information:

- Fatty liver *always* occurs if alcohol consumption exceeds 100 ml of ethanol each day (just over a bottle of wine). (Rachael was drinking three-quarters of a bottle of spirits, about twice this amount).

- The condition is usually reversible if no more alcohol is consumed.

- The condition may progress to cirrhosis and possibly liver failure if the person continues to drink alcohol (Teli et al., 1995).

For many people, indeed for Rachael herself, the use of alcohol is firmly embedded into social structures that are hard to dislodge. If Rachael stopped drinking how would she relieve the stresses and strains of her life? How could she remain at the core of her organisation when so much 'business' is conducted by her peers in the bar? If Rachael doesn't stop drinking, what might happen?

that people drift in and out of problem drinking over many years. Indeed only 22% reported abstinence for the entire 16 years.

Research findings collected from a large number of trials have also indicated which people can be predicted to have the best outcomes from various types of treatment for their problem drinking (Weisner et al., 2003.):

- Women seem to do better than men, although they often enter treatment later than men and may have more associated mental health problems.

- Although some young people 'mature' out of their problem drinking behaviour, in general very young people entering treatment have a worse outcome than more mature people.

- Lower socio-economic status is associated with less improvement over time.

- The worse the problem drinking at the start of treatment and the presence of psychiatric problems, the poorer the outlook for the success of treatment.

- Networks and family support can be important predictors of the outcome of treatment. If the person's family and friends are also involved in addictive types of behaviour, the chances of successful treatment are poorer. However, the presence of a supportive family environment is associated with improved chances of successful treatment (Copello et al., 2005).

5.6.2 Treating alcohol-related liver disorders

Although considerable progress has been made in the treatment of many other chronic medical conditions, scant progress has been made in the treatment of cirrhosis. In over 8000 people admitted to hospitals in the Oxford region of the UK with liver cirrhosis during a 30-year observation period, 34% had died one year after their admission and this death rate remained more or less constant (Roberts et al., 2005).

The largely pessimistic view of the failure of treatment of liver damage may change as there is evidence (Iredale, 2003) that the underlying processes, such as inflammation, are becoming increasingly understood, and may potentially be reversible in the future. However, it is clear that at the moment, no really effective medication currently exists to reverse the damage produced by alcohol and the best hope for long-term health for Rachael and other problem drinkers remains abstinence from drinking alcohol.

Liver transplantation

If Rachael continues to drink alcohol the fatty infiltration of her liver may progress to *cirrhosis* (Teli, 1999). When cirrhotic liver damage becomes severe, liver transplantation might be possible. Because of the limited availability of livers for transplantation some specialist units have a rule that people should demonstrate their resolve and not drink alcohol for a fixed period of time (perhaps 6 months), before the transplant is considered. This type of rule, with its rather moralistic overtones, has been challenged by Webb and Neuberger (2004). Other people may be not considered suitable for liver transplant because they have other physical or psychological problems (Walsh and Alexander, 2000).

Transplantation is a simple idea but replacing a diseased organ with a fully functioning one from a *donor* can be complicated in practice. As you can imagine there are many ethical dilemmas involved in taking organs from heart-beating but brain-dead people. Many of the potential medical problems arise after transplantation because the *host* person's body will consider the new organ to be 'non-self' and attempt to *reject* it. Although rejection can be controlled by immunosuppressive drugs, these often have side-effects. Despite these difficulties many people with a liver transplant lead entirely normal, active lives (Prasad and Lodge, 2001).

If Rachael's liver damage becomes so severe that she has to have a liver transplant then the outlook is surprisingly good. Approximately 75% of people with a new liver will be alive five years after the transplant surgery (Iredale, 2003), and progress with new surgical techniques and immunosuppressive drugs continues to improve the chances of survival. However, research has shown that between 8% and 22% of people are found to drink alcohol within 6 months of their transplant, and overall between 10% and 30% relapse in due course (Webb and Neuberger, 2004). Let us hope that Rachael is able to get the right information, treatment and support to help her live a long and happy life.

Summary of Chapter 5

5.1   The main acute effects of ethanol are on the nervous system, causing characteristic changes in behaviour and judgement. There are particular issues with regard to driving, with different countries setting various 'safe' limits for blood-ethanol concentration. Very high blood-ethanol concentrations can be fatal.

5.2   Hangovers are unpleasant and are poorly understood. Various mechanisms have been proposed including direct effects of ethanol on organs, ethanol withdrawal, accumulation of acetaldehyde and the effects of other chemicals present in alcoholic drinks. Many treatments are in common usage but there is little evidence of any particular intervention being beneficial.

5.3   Alcoholic liver disease results from excessive drinking and includes fatty liver (which is the early reversible stage) and the more serious alcohol-induced hepatitis and cirrhosis.

5.4   Excessive drinking can also lead to nervous system damage resulting in dementia, and shrinking of central nervous system tissue.

5.5   Fetal alcohol syndrome can result from maternal alcohol consumption during pregnancy. It involves disruption of fetal development causing CNS abnormalities, growth retardation and characteristic facial features.

5.6   Treatment of liver disorders is difficult other than by abstinence.

Learning outcomes for Chapter 5

After studying this chapter and its associated activities, you should be able to:

LO 5.1 Define and use in context, or recognise definitions and applications of, each of the terms printed in **bold** in the text. (Question 5.1)

LO 5.2 Describe the effects that ethanol in the blood has on the body – specifically the effect on drivers' judgement at blood-alcohol concentrations near the legal limit for driving and the serious effects of blood-alcohol concentrations in excess of 200 mg/100 ml. (Questions 5.1 and 5.5)

LO 5.3 Describe the short-term effects experienced following excessive alcohol consumption. Comment on the relationship between these complex effects and ethanol metabolites or congeners, the various 'remedies' adopted, and that the physiological basis for these economically important short-term harmful effects remains poorly understood. (Questions 5.2 to 5.5)

LO 5.4 Discuss some of the long-term harmful effects of drinking excessive alcohol with specific reference to the three stages of alcoholic liver disease and central and peripheral nervous system damage. (Questions 5.1 and 5.5)

LO 5.5 Discuss fetal alcohol syndrome, describing the main effects on the fetus and some of the factors that can make research into this disorder more complex. (Question 5.5)

Self-assessment questions for Chapter 5

Question 5.1 (LOs 5.1, 5.2 and 5.4)

Drinking alcohol produces a complex set of effects on a number of body systems. (a) On which system are the main acute effects most likely to lead to sudden death, and why? (b) Name some possible causes of death from drinking alcohol.

Question 5.2 (LO 5.3)

Explain two different ways whereby some individuals could have higher levels of acetaldehyde in their system than others, after drinking identical alcoholic drinks.

Question 5.3 (LO 5.3)

If the reaction catalysed by ALDH to form acetic acid is faster than the production of acetaldehyde (catalysed by ADH), how will this affect the drinker?

Question 5.4 (LO 5.3)

The proportion of fat per body weight increases with age. How might this affect older drinkers?

Question 5.5 (LOs 5.2 to 5.5)

(a) Why is it impossible to be precise about what is a *safe* drinking limit? (b) Why in particular are pregnant women advised to avoid drinking any alcohol at all?

CHAPTER 6

BALANCING HARMS WITH POSSIBLE BENEFITS

In this final chapter we turn to an area of some uncertainty about the effects of alcohol on human health. The evidence reviewed in preceding chapters points unequivocally to excessive alcohol consumption causing both acute and chronic physical and psychological harm. But is there a 'safe' level of alcohol drinking and could this moderate amount have a beneficial effect on health – at least for some individuals? You might have expected clear-cut answers to what seem simple questions, but there are a number of complex reasons why the issues are very difficult to resolve.

6.1   Is there a 'safe' level of alcohol intake?

As you already know, the level of alcohol consumption that is harmful varies with the individual and depends on several factors, including age, body mass, gender, previous medical history and genetic inheritance. The same is true for the level that is 'safe' to drink: there is no universal definition of how much alcohol a person can drink on a regular basis without suffering harm. Nevertheless, you will often see pronouncements in the media and from medical and government sources to the effect that 'moderate' consumption or 'drinking sensibly' does not cause harm as long as you don't drink and drive. The amounts quoted as 'moderate' tend to be 1 to 3 UK units per day, where a unit is the equivalent of around half a standard 175 ml glass of wine, or a 280–360 ml bottle or can of light beer, or a 20–30 ml measure of spirits. (The ranges given in these measures allow for variations in the % alcohol content by volume of different alcoholic drinks; see Chapter 1.)

However, generalisations about what constitutes a safe level are problematic: for example, the safe limit for a woman is on average lower than that for a man; people from certain ethnic groups can tolerate far less alcohol than the 'moderate' amount.

◆ How might drinking *patterns* influence the safe level of alcohol consumption?

◆ Even if two people consume the same moderate amount of alcohol every week, and are similar in other ways relevant to their health, there is much less risk associated with spreading the alcohol intake across the week than there is in 'bingeing' it all in one drinking session (Chapter 1).

Age is another important factor: teenagers and young adults are more likely than older adults to save their drinking for one big night out every week. But consuming 14 'drinks' in one evening is far more damaging than consuming two drinks a day for a week, which is within 'drinking safely' guidelines.

6.2   Can moderate drinking bring any health benefits?

From the 1920s onwards, evidence has been slowly building that moderate amounts of alcohol taken regularly might have some protective effect against certain disease states. Large population studies (reviewed by Marmot, 2001; Bovet and Paccaud, 2001) have shown that moderate drinkers tend to have lower mortality rates from certain causes than people who drink little or no alcohol, and that drinking moderately is associated with at least a 20% reduction in the risk of developing **coronary heart disease (CHD)**. This condition occurs when the arteries supplying oxygen and nutrients to the heart muscle become blocked by fatty deposits known as *plaques* (pronounced 'plax', Figure 6.1), and areas of heart muscle die as a result. Also, as Chapter 1 pointed out, if everyone in developed countries *stopped* drinking alcohol, there would be about 17% more cases of ischaemic stroke (WHO, 2002) – brain damage caused when arteries in the brain are blocked by plaque formation or clots.

The problem with defining safe limits is that drinking even a moderate amount of alcohol probably causes some harms (for example, one drink a day increases blood pressure), as well as some benefits, so a 'trade-off' is occurring within each individual. The best that anyone can do is make an informed guess about whether the benefits are likely to outweigh the harms in their own case. Coronary heart disease illustrates this very well.

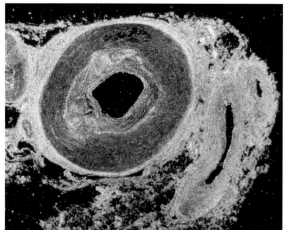

Figure 6.1 The space within this coronary artery has been greatly reduce by hardened fatty deposits (plaques) on its inner wall, which restrict the amount of blood that can supply the heart muscle. Plaques may also break off and block the circulation of blood to the heart, triggering a heart attack. (Photo: Science Photo Library)

6.2.1 Protection from coronary heart disease

Biological research has established several plausible mechanisms for the reduction in CHD risk associated with drinking alcohol moderately over several years. Ethanol in the bloodstream seems to counteract the accumulation of factors that contribute to the development of plaques and it reduces clot formation, so the coronary arteries are less likely to become blocked.

However, a reduction in the risk of CHD is only evident in people *who were at significant risk of developing it*. This point is crucial: moderate alcohol drinking in young people and in countries where CHD risk is very low (e.g. in Africa) and in women before the menopause, has no protective effect against CHD because people in these categories are at such a low risk of developing it. In these groups, whatever harms may be associated with moderate alcohol consumption (e.g. from binge drinking, increased risk of road traffic accidents, or greater likelihood of progressing to heavy drinking later in life) are not counterbalanced by any protective effect of alcohol against CHD (Figure 6.2).

Guidelines about beneficial levels of alcohol are stated cautiously, in part because population studies can't prove that an association between two variables (in this case the amount of alcohol consumed and the incidence of, or mortality from, CHD) is *causal*, i.e. they cannot prove that moderate alcohol intake *causes* a reduction in CHD. It could be that the direction of causality is the *opposite* of the one you might expect: perhaps there is something about healthier people (who are less likely to develop CHD) which makes them also more likely to drink moderately or less likely to drink excessively?

Figure 6.2   The likely benefits of moderate alcohol drinking may not outweigh the potential harms except in some men aged over 40 years and some women over 50. (Source: Assunta Del Buono/John Birdsall Photo Library)

◆ Does uncertainty about the direction of causality, as described above, suggest a possible explanation for the higher mortality rates from CHD in people who don't drink alcohol at all, compared with the moderate drinkers?

◆ Instead of abstinence causing a higher risk of CHD, perhaps it's the other way around? The symptoms of early CHD might cause people to give up drinking alcohol, i.e. they are abstinent because they didn't feel well enough to drink, or because they were advised to give up drinking.

Another complication is illustrated by studies which show that wine drinkers experience a greater protective effect against CHD than people who drink similar amounts of alcohol in beer or spirits (cited in Bovet and Paccaud, 2001). This suggests that there could be something other than the alcohol in wine which gives it a greater protective effect (more on this later).

◆ Can you suggest an alternative explanation?

◆ You may have wondered whether wine-drinkers are, in general, more affluent than beer-drinkers, and perhaps more able to afford healthier foods, have more opportunities to exercise, take holidays and enjoy greater access to health care. In other words, their wine drinking may simply be an indicator of their greater affluence, which enables them to afford the *real* underlying causes of their better health.

Wine drinking in the above scenario is an example of a **confounding factor**, a concept of huge importance in the interpretation of health data. At first sight, the data seem to be saying that wine drinking could cause a reduction in CHD, but on closer inspection it may just be an indicator of the genuine cause or causes. A confounding factor in examples like this does not cause the better outcome, but it is associated with better health solely *because* it is also associated with the true cause of better health.

Even with all the caveats outlined above, there is a consensus that *at the population level* there is some protection against CHD and ischaemic stroke from drinking alcohol moderately and regularly over a long period, particularly with meals, which slows the rate of ethanol absorption. But there is no way of telling whether the benefits will outweigh the possible harms in *your* case.

6.2.2 Other possible health benefits

The evidence for a protective effect from moderate alcohol intake against other health conditions is far less secure. A few large-scale population studies have found some reduction in the risk of developing *dementia*, including Alzheimer's disease, after the age of 60 years in people who drink light to moderate amounts of alcohol daily (Ruitenberg et al., 2002; Wright, 2006). The mechanism may be similar to that for ischaemic stroke in reducing brain damage due to insufficient blood supply (Stampfer, 2006). There is also some evidence that a moderate intake of alcohol improves *bone density* and bone strength in older people and may result in a reduced risk of fractures (Jugdaohsingh et al., 2006).

Bone density and fractures are discussed in detail in another book in this series (Phillips, 2008).

Earlier we mentioned the possibility that something in wine *other than* the alcohol might have a protective effect against some conditions. The most interesting lines of inquiry suggest that a chemical derived from the skins and seeds of grapes, which is found in red wine, may reduce the risk of developing certain cancers (National Cancer Institute, 2007). It may not be the alcohol content that is responsible: eating red fruits could be just as beneficial.

So far we have only referred to the direct actions on the body of one or more constituents of alcoholic drinks. However, it should not be forgotten that most people who drink moderately often do so in social settings, including outside the home. In many societies alcohol is often involved in facilitating social interactions with other people, which can bring their own health benefits by reducing loneliness and counteracting depression.

6.3 Final comments

In this book, you have explored many different aspects of alcohol, including its chemical effects on the body and the consequences of excessive drinking for health. For many people it is a relaxing and enjoyable feature of social occasions; for others it has religious significance and may be venerated or forbidden. You have seen evidence of the serious consequences of alcohol abuse in the mortality data from traffic accidents, liver cirrhosis and alcoholic poisoning, followed Rachael's story as alcohol gradually took over her life and damaged her career, her financial security and her relationships as well as her health and listened to the real accounts of three recovering alcoholics. You now know that there are no definitive guidelines on what might constitute a 'safe' level of alcohol consumption for any individual, but you may be interested to know that some of the authors of this book still enjoy a glass or two from time to time!

Summary of Chapter 6

6.1   Drinking moderate amounts of alcohol regularly over a long period from middle-age onwards is associated with a reduced risk of coronary heart disease and ischaemic stroke, and possibly also protects against the risk of dementia, and fractures due to bone thinning in later life.

6.2   Moderate alcohol consumption is generally stated as 1 to 3 UK alcohol units per day; certain health risks appear to be greater above and below this level.

6.3   Population studies showing associations between alcohol intake and health outcomes must be interpreted with caution: the direction of causality may be the opposite of what you intuitively expect; alcohol may turn out to be a confounding factor, which is independently associated with the real cause of the health effect. The trade-off between possible harms and possible benefits cannot be estimated accurately for individuals.

Learning outcomes for Chapter 6

After studying this chapter and its associated activities, you should be able to:

LO 6.1   Define and use in context, or recognise definitions and applications of, each of the terms printed in **bold** in the text. (Questions 6.1 and 6.2)

LO 6.2   Summarise the evidence on the possible beneficial effects of moderate alcohol consumption. (Questions 6.1 and 6.2)

LO 6.3   Explain why population studies showing an association between moderate alcohol intake and a reduced risk of developing certain disease states cannot be assumed to prove that alcohol is causing the improved health outcome. (Questions 6.1 and 6.2)

Self-assessment questions for Chapter 6

Question 6.1 (LOs 6.1 to 6.3)

In Chapter 1, Table 1.7 showed that worldwide, in the year 2000, the number of disability adjusted life years (DALYs) attributable to alcohol-related cardiovascular disease (heart disease, strokes, and high blood pressure) exceeded 4.4 million in men, but the value for women was *minus* 428 thousand DALYs. Explain this paradox.

Question 6.2 (LOs 6.1 to 6.3)

Based on the discussion given in this chapter, is alcohol a confounding factor in studies that show drinking red wine regularly is associated with a reduced risk of some cancers? Explain your answer.

ANSWERS TO SELF-ASSESSMENT QUESTIONS

Question 1.1

If the total amount consumed in a population is divided by the total number of people, *including* the adult abstainers and those who are too young to drink, then the estimated average amount per person will *under*estimate the amount consumed on average by alcohol drinkers.

Question 1.2

40% of the 25 ml of whisky in the glass is pure alcohol. So the amount of pure alcohol is:

$$\frac{40}{100} \times 25 \text{ ml} = 10 \text{ ml of pure alcohol}$$

One litre (1 l) = one thousand millilitres (1000 ml) or one hundred centilitres (100 cl). So there are ten millilitres (10 ml) in every centilitre. To convert 10 ml of pure alcohol into centilitres you divide by 10, so the answer is 1 cl of alcohol.

Question 1.3

The claim is correct. The lager strength is 4% vol., therefore 500 ml contains:

$$\frac{4}{100} \times 500 \text{ ml} = 20 \text{ ml alcohol}$$

1 UK unit of alcohol is 10 ml, therefore the can contains 2 units.

Question 1.4

Although the average amount consumed in most of Western Europe is amongst the highest in the world, the dominant drinking pattern of 'wine with meals' spreads this consumption across most days in the year. The amount consumed on any occasion is relatively small and high-risk binge drinking to achieve intoxication is uncommon.

Question 1.5

(a) The countries with the highest self-reported proportion of drunkenness among 15-year olds tend to be in Northern and Eastern Europe, whereas the lowest rates are in Southern European countries.

(b) A higher proportion of boys than girls in every country report being drunk at least 10 times in the previous year, but the difference is very small in Iceland and Norway.

(c) More than a quarter of 15-year-olds in the top four countries – which include the UK and Ireland – say they are getting drunk 10 times or more in a year.

Question 1.6

Young people are more likely than older adults to be involved in alcohol-related behaviours that result in accidental or violent injuries or death. Disabilities and deaths occurring at a young age weigh heavily in estimates of DALYs, which take into account the years of healthy life lost. The chronic alcohol-related diseases that affect people who have been drinking to excess for a long time result in *fewer* years of life lost to disability or premature death, because their impact occurs at much older ages.

Question 1.7

The USA and Argentina both have alcohol consumption in the same band (Figure 1.3), but despite this similarity you cannot assume that they have identical proportions of males in each age-group. It would be advisable to age-standardise the death rates before you compare them, to ensure that the data are not distorted by different proportions of males in their 40s and 50s – the peak age for alcoholic poisoning.

Question 1.8

The acute effects of alcohol include traffic accidents, injuries, violence and alcoholic poisoning, which place a major drain on police resources and the emergency medical services, as well as disrupting the ability of those affected to earn an income, pay taxes and spend on goods and services. Chronic diseases and disorders due to excessive alcohol consumption lead to significant reductions in productivity, increased payout of sickness or other benefits and major health-care costs.

Question 2.1

In N_2, each N atom shares three electrons to form a triple bond, $N{\equiv}N$. In ammonia, NH_3, a single bond between the nitrogen and each hydrogen atom, is formed:

$$
\begin{array}{c}
H \\
| \\
H\!-\!N\!-\!H
\end{array}
$$

Question 2.2

If calcium loses two electrons, the charge on the ion will be 2+. This is written Ca^{2+}. To balance the 2+ charge on calcium, two chloride ions, Cl^-, will be needed. The formula of calcium chloride is thus $CaCl_2$.

Question 2.3

You know that each oxygen shares two electrons, so in this molecule it forms double bonds to sulfur:

$$O = S = O$$

Question 2.4

$$
\begin{array}{c}
\text{H} \quad \text{O} \\
| \quad \parallel \\
\text{H---C---C} \\
| \quad \backslash \\
\text{H} \quad \text{H}
\end{array}
\quad \text{and} \quad
\begin{array}{c}
\text{H} \quad \text{O} \\
| \quad \parallel \\
\text{H---C---C} \\
| \quad \backslash \\
\text{H} \quad \text{O---H}
\end{array}
$$

Question 2.5

The ball-and-stick model shows the position of the atoms so that the distances between atoms and the geometry of the surrounding bonds are clearly visible. The space-filling model gives a more realistic view of the real shape of the molecule, including the area occupied by the outer electrons.

Question 2.6

The hydrogen atoms of H_2O, with their positive charge are attracted to the negatively charged bromide ion, so water molecules surround the Br^- ion; conversely the negatively charged oxygen atoms of H_2O surround the K^+ ion. The two ions are now effectively separated from each other by the water molecules, and can act independently – this is the process of dissolving to form an aqueous solution.

Question 2.7

When calcium iodide in the solid state (s) dissolves, it forms one calcium ion (Ca^{2+}) and two iodide ions ($2I^-$) in aqueous solution (aq).

$$CaI_2(s) = Ca^{2+}(aq) + 2I^-(aq)$$

superscript, prefix, subscript, state

Question 2.8

(a) Hydrogen gas, H_2, reacts with chlorine gas, Cl_2, to form gaseous hydrogen chloride, HCl. The unbalanced equation is:

$$H_2 + Cl_2 \longrightarrow HCl \tag{2.18}$$

which does not balance as there are two hydrogens on the left and only one on the right; similarly for chlorine. Starting therefore by balancing the numbers of hydrogen atoms, add a prefix 2 to the right-hand side to make two molecules of HCl.

$$H_2 + Cl_2 \longrightarrow 2HCl \tag{2.19}$$

Now balance chlorine. In Equation 2.7, there are two chlorines on the left and two on the right – so the numbers of chlorine atoms balance as well. Finally, put the finishing touches – the equals sign and the physical states of the molecules.

$$H_2(g) + Cl_2(g) = 2HCl(g) \tag{2.20}$$

(b) $3H_2(g) + N_2(g) = 2NH_3(g)$

(c) $CaCl_2(s) = Ca^{2+}(aq) + 2Cl^-(aq)$.

Be careful here – a chlorine atom can only gain one electron so that the chloride ion always carries a single negative charge; thus *two* chloride ions are formed as indicated by the prefix 2, balanced by *one* calcium with two positive charges.

Question 2.9

(a) $2C_4H_{10}(g) + 13O_2(g) = 8CO_2(g) + 10H_2O(g)$

The reason for taking two molecules of butane in the equation is that it prevents balancing the equation with 6½ molecules of oxygen.

(b) $C_6H_{12}O_6(aq) = 2C_2H_5OH(aq) + 2CO_2(g)$

Question 2.10

The energy given out by the burning of methane is calculated from the following equation.

$$CH_4(g) + 2O_2(g) = CO_2(g) + 2H_2O(g) \tag{2.10}$$

Methane has four C$-$H bonds only.

type of bond broken	number of bonds broken	bond energy (kJ/mol)	total energy of bonds broken	type of bond formed	number of bonds formed	bond energy (kJ/mol)	total energy of bonds formed
C$-$H	4	413	1652	C$=$O	2	770	−1540
O$=$O	2	498	996	O$-$H	4	464	−1856
sub total			**2648**				**−3396**

The energy released in this reaction is therefore $+2648 - 3396$ kJ/mol = -748 kJ/mol.

Question 3.1

The presence of food in the stomach causes the pyloric sphincter to close, in order to hold the food in the stomach until it is sufficiently digested to be released into the small intestine. If the stomach is empty then the pyloric sphincter is open. Therefore if an alcoholic drink is consumed when the stomach is empty it will pass straight from the stomach into the small intestine where it will be rapidly absorbed into the bloodstream. This will result in faster intoxication than when food is present and the pyloric sphincter is closed (in that case alcohol is held in the stomach for longer and released slowly into the small intestine, delaying absorption and subsequent intoxication).

Question 3.2

Ethanol in the bloodstream causes widening of the blood vessels (vasodilation) in the skin. When these vessels dilate, they carry more blood so the skin turns a red colour and gets warm (this is a natural mechanism for cooling the body).

Question 3.3

The liver may be damaged by alcohol consumption more than other organs because it receives a higher blood-alcohol (ethanol) concentration. This is because when ethanol is absorbed through the wall of the gut it is transported directly to the liver in the portal vein. The liver receives three-quarters of its blood in this way (the rest comes from the normal circulation via the hepatic artery) whereas other organs receive all of their blood from the normal circulation (which has a lower ethanol concentration). The liver removes some ethanol from the blood, so the blood leaving the liver and travelling to the other organs will always have a lower ethanol concentration than the blood travelling to the liver in the portal vein.

Question 3.4

Ethanol in the blood inhibits the release of vasopressin, a hormone which is normally secreted from the brain in response to decreased water levels in the body. Because vasopressin acts in the kidneys by causing the cells that line the nephron tubules to move more water back into the bloodstream, a decrease in vasopressin (as caused by consumption of alcohol) will reduce this water movement, so more water will remain in the tubules and become urine.

Question 4.1

The claim could mean that at least part of the effect could be due, not to the chemical properties of ethanol, but to the context in which it arrives in the body. For example, the presence of alcohol-related conditional stimuli and people's expectations that they are drinking alcohol could make them behave with fewer inhibitions, as though they had been drinking alcohol.

Question 4.2

(ii) An antagonist to dopamine. Based on the understanding described, dopamine is a neurotransmitter that underlies the seeking/attraction of alcohol. Therefore, an *agonist* would tend to enhance alcohol seeking whereas an *antagonist* would tend to block it.

Question 4.3

Rachael would seem to meet all the criteria for addiction in Table 4.1, at least to some extent. 1. *Tolerance*: over time she needed increasing amounts of alcohol to achieve the same effect. 2. *Withdrawal*: if she stopped drinking for a few days she experienced withdrawal symptoms. 3. *Loss of control*: she fell down stairs when drunk at university and had unprotected sex with men she hardly knew. 4. *Attempts to control*: she always made a New Year's resolution to stop drinking, but never kept it. 5. *Time spent on use*: she had to rearrange her diary sometimes in order to sober up and at university she missed a lot of lectures. 6. *Sacrifices made for use*: this isn't clear, but her drinking was a financial drain on the family budget, so she may have given up other activities to compensate. 7. *Use despite known suffering*: alcohol has blunted Rachael's cognitive abilities and she has suffered many hangovers and feelings of guilt; in Chapter 5 you will discover how it has damaged her health.

Question 5.1

(a) The main acute effects are on the nervous system, causing mood changes and impairment of judgement and reaction time, etc. (b) Possible causes of death are a fatal traffic accident due to poor judgement; other accident due to uninhibited behaviour; inhalation of vomit; heart or lung failure; liver cirrhosis; suicide due to depression. You may be able to think of others.

Question 5.2

(a) Acetaldehyde is produced from ethanol more rapidly in some individuals than in others.

(b) Acetaldehyde is converted into acetic acid more slowly in some individuals than in others.

Question 5.3

There will be no build-up of acetaldehyde because it will metabolise to acetic acid as soon as it is made. So, the drinker will not experience the flushing, increased heart rate, dizziness or nausea associated with a build-up of acetaldehyde.

Question 5.4

The increased proportion of fat to muscle in older people will result in a decrease in total body water. As ethanol is water-soluble, the same amount of ethanol will be dissolved in a smaller amount of water, resulting in a higher BAC in an older person than a younger. Unless the body responds by making more enzymes, this will result in higher levels of intoxication after smaller amounts of ethanol.

Question 5.5

(a) People inherit genes that direct production of ethanol-metabolising enzymes, and will therefore process ethanol at different rates. People also vary greatly in weight and in their muscle to fat ratios. It is therefore impossible to do more than give very broad guidelines on what might be a safe level of drinking. (b) Pregnant women are advised to abstain from drinking alcohol because of the risk of fetal alcohol syndrome (FAS). No safe limit has been identified for this because the mechanisms by which alcohol consumption causes FAS in some individuals are poorly understood.

Question 6.1

A trade-off between alcohol-related harms and benefits is occurring. In the male population of the world, on average the harms outweigh the benefits, because they are more likely than women to drink heavily and to binge drink. If men reduced their alcohol consumption worldwide, then the total number of DALYs attributable to alcohol-related cardiovascular disease would be *less* than 4.4 million. In the world's female population, on average the benefits outweigh the harms; moderate drinking has a protective effect against cardiovascular disease and 428 thousand *more* DALYs are prevented than are caused by drinking alcohol.

Question 6.2

Alcohol *is* a confounding factor in these studies. The discussion in the chapter states that it does not cause the reduction in some cancers, which is probably due to chemicals in the wine derived from grape seeds and skins. Alcohol is merely *associated* with the reduction because it is a constituent of red wine, but you might easily draw the incorrect conclusion from the results that alcohol itself is having a protective effect.

REFERENCES AND FURTHER READING

References

Academy of Medical Sciences (2004) *Calling Time: The Nation's Drinking as a Major Health Issue*, London, The Academy of Medical Sciences. Available from: http://www.acmedsci.ac.uk/images/publication/pcalling.pdf (Accessed May 2006)

Alcohol Concern (2002) *Report on the Mapping of Alcohol Services in England*, London, Alcohol Concern.

Alcohol Concern Briefings and Fact Sheets (2005) [online] Young people's drinking; Drink-drive accidents; Suicide and alcohol misuse (Briefing 5: Mental Health and Alcohol Misuse Project). Available from: http://www.alcoholconcern.org.uk/servlets/wrapper/fact_sheets.jsp (Accessed May 2007)

American Psychiatric Association (1994) *Diagnostic and Statistical Manual of Mental Disorders*, 4th edn, Washington, DC, American Psychiatric Association.

Arnon. R., Degli Esposti, S. and Zern, M. A. (1995) 'Molecular biological aspects of alcohol-induced liver disease,' *Alcoholism: Clinical and Experimental Research*, vol. 19, pp. 247–256.

Becker, U., Deis, A., Sørensen, T. I. A., Grønbæk, M., Borch-Johnsen, K., Müller, C. F., Schnohr, P. and Jensen, G. (1996) 'Prediction of risk of liver disease by alcohol intake, sex, and age: A prospective population study', *Hepatology*, vol. 23, pp. 1025–1029.

Berridge, K. C. (2003) 'Pleasures of the brain', *Brain and Cognition*, vol. 52, pp. 106–128.

Bloomfield, K., Stockwell, T., Gmel, G. and Rehn, N. (2003) 'International comparisons of alcohol consumption', *Alcohol Research and Health*, vol. 27, pp. 95–109.

Bonomo, Y. and Proimos, J. (2005) 'Substance misuse: alcohol, tobacco, inhalants and other drugs', *British Medical Journal*, vol. 330, pp. 770–780.

Boreham, R. and Shaw, A. (2002) *Smoking, Drinking and Drug Use among Young People*, London, Stationery Office.

Bovet, P. and Paccaud, F. (2001) 'Commentary: Alcohol, coronary heart disease and public health: which evidence-based policy?' *International Journal of Epidemiology*, vol. 30, pp. 734–737.

Brand-Miller, J. C., Fatima, K., Middlemiss, C., Bare, M., Liu, V., Atkinson, F. and Petocz, P. (2007) 'Effect of alcoholic beverages on postprandial glycemia and insulinemia in lean, young, healthy adults', *American Journal of Clinical Nutrition*, vol. 85, pp. 1545–1551.

Buonopane, A. and Petrakis, I. (2005) 'Pharmacology of alcohol use disorders', *Substance Use and Misuse*, vol. 40, pp. 2001–2020.

Cabinet Office (2003) *Alcohol Misuse: How much does it cost?* London, Stationery Office. Available from: http://sia.dfc.unifi.it/costi%20uk.pdf (Accessed April 2006)

Cabinet Office Strategy Unit (2004) *Alcohol Harm Reduction Strategy for England*, London, Strategy Unit. Available from: http://www.strategy.gov.uk/downloads/su/alcohol/pdf/CabOffce%20AlcoholHar.pdf (Accessed April 2006)

Cooper, M. L. (2006) 'Does drinking promote risky sexual behaviour?' *Current Directions in Psychological Science*, vol. 15, pp. 19–23.

Copello, A., Velleman, R. and Templeton, L. (2005) 'Family interventions in the treatment of alcohol and drug problems', *Drug and Alcohol Review*, vol. 24, pp. 368–385.

Dhital, R. et al. (2001) *Alcohol and Drugs in Nepal: With Reference to Children*. Kathmandu: Child Workers in Nepal Concerned Centre (CWIN); cited in WHO (2004) *Global Status Report on Alcohol*, Geneva: World Health Organization.

Ferrari, P., Jenab, M., Norat, T. et al. (2007) 'Lifetime and baseline alcohol intake and risk of colon and rectal cancers in the European prospective investigation into cancer and nutrition (EPIC)' *International Journal of Cancer*. Available from: http://www3.interscience.wiley.com/cgi-bin/fulltext/114294429/HTMLSTART (Accessed August 2007)

Field, M., Mogg, K. and Bradley, B. P. (2005) 'Alcohol increases cognitive biases for smoking cues in smokers', *Psychopharmacology*, vol. 180, pp. 63–72.

Foster, S. E., Vaughan, R. D., Foster, W. H. and Califano, J. A. (2006) 'Estimate of the commercial value of underage drinking and adult abusive and dependent drinking to the alcohol industry', *Archives of Pediatric and Adolescent Medicine*, vol. 160, pp. 473–478.

Garfield, C. F., Chung, P. J. and Rathouz, P. J. (2003) 'Alcohol advertising in magazines and adolescent readership', *Journal of the American Medical Association*, vol. 289, pp. 2424–2429.

Halliday, T. R. and Davey, G. C. B. (eds) (2007) *Water and Health in an Overcrowded World*, Oxford, Oxford University Press.

Hibell, B., Andersson, B., Ahlström, S., Balakireva, O., Bjarnason, T., Kokkevi, A. and Morgan, M. (2000) *The 1999 ESPAD Report. Alcohol and Other Drug Use Among Students in 30 European Countries*, The Swedish Council for Information on Alcohol and Other Drugs (CAN) and The Pompidou Group at the Council of Europe, Stockholm, Sweden. Available from: http://www.espad.org/sa/node.asp?node=641 (Accessed May 2007)

Hingson, R., Heeren, T., Zakocs, R., Winter, M. and Wechsler, H. (2003) 'Age of first intoxication, heavy drinking and risk of unintentional injury among US college students', *Journal of Studies on Alcohol*, vol. 64, pp. 23–31.

Hull, J. G., Young, R. D. and Jouriles, E. (1986) 'Applications of the self-awareness model of alcohol consumption – predicting patterns of use and abuse', *Journal of Personality and Social Psychology*, vol. 51, pp. 790–796.

Iredale, J. (2003) 'Cirrhosis: new research provides a basis for rational and targeted treatments', *British Medical Journal*, vol. 327, pp. 143–147.

Jones A. W. (1987) 'Elimination half-life of methanol during hangover', *Pharmacology and Toxicology*, vol. 60, pp. 217–220.

Jones, B. T., Jones, B. C., Thomas, A. P. and Piper, J. (2003) 'Alcohol consumption increases attractiveness ratings of opposite-sex faces: a possible third route to risky sex', *Addiction*, vol. 98, pp. 1069–1075.

Josephs, R. A. and Steele, C. M. (1990) 'The two faces of alcohol myopia: Attentional mediation of psychological stress', *Journal of Abnormal Psychology*, vol. 99, pp. 115–126.

Jugdaohsingh, R., O'Connell, M. A., Sripanyakorn, S., Powell, J. J. (2006) 'Moderate alcohol consumption and increased bone mineral density: potential ethanol and non-ethanol mechanisms', *Proceedings of the Nutrition Society*, vol. 65, pp. 291–310.

Klingemann, H. and Sobell, L. (2001) 'Introduction: natural recovery research across substance use', *Substance Use and Misuse*, vol. 36, pp. 1409–1416.

Koob, G. F. and Le Moal, M. (2006) *Neurobiology of Addiction*, Amsterdam, Elsevier.

Leifman, H. (2001) 'Estimations of unrecorded alcohol consumption levels and trends in 14 European countries', *Nordic Studies on Alcohol and Drugs*, vol. 18 (English Supplement), pp. 54–70; cited in *'Calling Time: The Nation's Drinking as a Major Health Issue'*, A report from the Academy of Medical Sciences, March 2004, UK.

Leon, D. and McCambridge, J. (2006) 'Liver cirrhosis mortality rates in Britain from 1950 to 2002: an analysis of routine data', *The Lancet*, vol. 367, pp. 52–56.

Lynn, M. (1988) 'The effects of alcohol consumption on restaurant tipping', *Personality and Social Psychology Bulletin*, vol. 14, pp. 87–91.

Lyvers, M. (2000) 'Loss of 'control' in alcoholism and drug addiction: A neuroscientific interpretation', *Experimental and Clinical Psychopharmacology*, vol. 8, pp. 225–249.

Mann, K., Schafer, D., Ackermann, K. and Croissant, B. (2005) 'The long-term course of alcoholism, 5, 10 and 16 years after treatment', *Addiction,* vol. 100, pp. 797–805.

Marmot, M. G. (2001) 'Commentary: reflections on alcohol and coronary heart disease', *International Journal of Epidemiology*, vol. 30, pp. 729–734.

Mental Health Foundation (2006) *Cheers! Understanding the Relationship between Alcohol and Mental Health*, London, The Mental Health Foundation.

Midgley, C. A. (ed.) (2008) *Chronic Obstructive Pulmonary Disease: A Forgotten Killer*, Oxford, Oxford University Press, in press.

National Cancer Institute (2007) [online] Red wine and cancer prevention: Fact Sheet. Available from: http://www.cancer.gov/cancertopics/factsheet/red-wine-and-cancer-prevention (Accessed February 2007)

National Institutes of Health (2003) [online] Understanding alcohol. Available from: http://science.education.nih.gov/supplements/nih3/alcohol/guide/info-alcohol.htm (Accessed May 2007)

National Institute on Alcohol Abuse and Alcoholism (2000) [online] Alcohol Alert 50: Fetal Alcohol Exposure and the Brain. Available from: http://pubs.niaaa.nih.gov/publications/aa50.htm (Accessed May 2007).

Nemtsov, A. (2005) 'Russia: alcohol yesterday and today', *Addiction*, vol. 100, pp. 146–149.

NERA (2003) *Alcohol in London: a cost-benefit analysis*, London, National Economic Research Associates.

Nesse, R. M. and Berridge, K. C. (1997) 'Psychoactive drug use in evolutionary perspective', *Science*, vol. 278, pp. 63–66.

Parvin, E. M. (ed.) (2007) *Screening for Breast Cancer*, Oxford, Oxford University Press, in press.

Paton, A. (2005) 'ABC of alcohol: Alcohol in the body', *British Medical Journal*, vol. 330, pp. 85–87.

Phillips, J. B. (ed.) (2008) *Trauma, Repair and Recovery*, Oxford, Oxford University Press, in press.

Pittler, M. H., Verster, J. C. and Ernst, E. (2005), Interventions for preventing or treating alcohol hangover: systematic review of randomised controlled trials. *British Medical Journal*, vol. 331, pp. 1515–1518.

Prasad, K. and Lodge, J. (2001) 'Transplantation of the liver and pancreas', *British Medical Journal*, vol. 322, pp. 845–847.

Pridemore, W. A. and Kim, S-W. (2006) 'Research note: Patterns of alcohol-related mortality in Russia', *Journal of Drug Issues*, vol. 36, pp. 229–247.

Prime Minister's Strategy Unit (2004) *Alcohol Harm Reduction Strategy for England*, Cabinet Office.

Reuben, A. (2006) 'Alcohol and the liver', *Current Opinion in Gastroenterology*, vol. 22, pp. 263–271.

Roberts, S., Goldacre, M. and Yeates, D. (2005) 'Trends in mortality after hospital admission for liver cirrhosis in an English population from 1968 to 1999', *Gut*, vol. 54, pp. 1615–1621.

Robinson, T. E. and Berridge, K. C. (1993) 'The neural basis of drug craving: an incentive-sensitization theory of addiction', *Brain Research Reviews*, vol. 18, pp. 247–291.

Ruitenberg, A., van Swieten, J. C., Witteman, J. C. M., Mehta, K. M., van Duijn, C. M., Hofman, A., Breteler, M. M. B. (2002) Alcohol consumption and risk of dementia: the Rotterdam Study, *The Lancet*, vol. 359, pp. 281–286.

Safe Travel (2006) [online] Available from: http://www.safetravel.co.uk/ (Accessed May 2007)

Soyka, M. and Roesner, S. (2006) 'New pharmacological approaches for the treatment of alcoholism', *Expert Opinion on Pharmacotherapy*, vol. 7, pp. 2341–2353.

Stampfer, M. J. (2006) 'Cardiovascular disease and Alzheimer's disease', *Journal of Internal Medicine*, vol. 260, pp. 211–223.

Steele, C. M. and Josephs, R. A. (1988) 'Drinking your troubles away. 2. An attention-allocation model of alcohol's effect on psychological stress', *Journal of Abnormal Psychology*, vol. 97, pp. 196–205.

Strömland, K. (2004) 'Fetal alcohol syndrome – a birth defect recognized worldwide', *Fetal and Maternal Medicine Review*, vol. 15, pp. 59–71.

Tapson, F. (2004) [online] The alcohol content of drinks. Available from: http://www.cleavebooks.co.uk/dictunit/notes6.htm (Accessed May 2007)

Teli, M., Day, C. and Burt, A. (1999) 'Determinants of progression to cirrhosis or fibrosis in pure alcoholic fatty liver.' *The Lancet*, vol. 346, pp. 987–990.

Tiffany, S. T. (1990) 'A cognitive model of drug urges and drug-use behavior: Role of automatic and nonautomatic processes', *Psychological Review*, vol. 97, pp. 147–168.

Toates, F. (ed.) (2007) *Pain*, Oxford, Oxford University Press.

Tomie, A. (1996) 'Locating drug reward cue at response manipulandum (CAM) induces symptoms of drug abuse', *Neuroscience and Biobehavioral Reviews*, vol. 20, pp. 505–535.

Walsh, K. and Alexander, G. (2000) 'Alcoholic liver disease', *Postgraduate Medical Journal*, vol. 76, pp. 280–286.

Webb, K. and Neuberger, J. (2004) 'Transplantation for alcoholic liver disease', *British Medical Journal*, vol. 329, pp. 63–64.

Weisner, C., Matzger, H. and Kaskutas, L. (2003) 'How important is treatment? One year outcomes of treated and untreated alcohol-dependent individuals', *Addiction,* vol. 98, pp. 901–911.

Weiss, F. (2005) 'Neurobiology of craving, conditioned reward and relapse', *Current Opinion in Pharmacology*, vol. 5, pp. 9–19.

WHO (2002) *The World Health Report 2002: Reducing Risks, Promoting Healthy Life*, Geneva, World Health Organization.

WHO (2004) *Global Status Report on Alcohol*, Geneva, World Health Organization.

WHO (2005a) *Public Health Problems Caused by Harmful Use of Alcohol*, Geneva, World Health Organization.

WHO (2005b) *Alcohol policy in the WHO European region: current status and the way forward*, Fact Sheet EURO/10/05, Copenhagen, Bucharest, World Health Organization.

WHO (2006) [online] *Global Alcohol Database*. Available from: http://www3.who.int/whosis/menu.cfm?path=whosis,alcohol&language=english (Accessed August 2007)

Wright, C. B., Elkind, M. S. V., Luo, X. D., Paik, M. C., Sacco, R. L. (2006) 'Reported alcohol consumption and cognitive decline: The Northern Manhattan Study', *Neuroepidemiology*, vol. 27, pp. 201–207.

Zeichner, A. and Phil, R. O. (1979) 'Effects of alcohol and behavior contingencies on human aggression', *Journal of Abnormal Psychology*, vol. 88, pp. 153–160.

Further reading

Paton, A. and Touquet, R. (eds) (2005) *ABC of Alcohol*, 4th edn, Oxford, BMJ Books, Blackwell Publishing.

Useful websites, maintained by the OU Library through the ROUTES system:

NHS Direct

http://www.nhsdirect.nhs.uk/articles/article.aspx?articleId=846

National Institute on Alcohol Abuse and Alcoholism

http://www.niaaa.nih.gov/

http://science.education.nih.gov/supplements/nih3/alcohol/guide/info-alcohol.htm

Alcohol Concern

http://www.alcoholconcern.org.uk/servlets/home

How Stuff Works

http://www.howstuffworks.com/alcohol.htm

http://health.howstuffworks.com/hangover.htm

http://health.howstuffworks.com/alcoholism.htm

chemcases.com

http://chemcases.com/alcohol/index.htm

Cleave books/Frank Tapson calculators

http://www.cleavebooks.co.uk/dictunit/notes6.htm

Alcoholics Anonymous

http://www.alcoholics-anonymous.org.uk/

http://www.alcoholics-anonymous.org/?Media=PlayFlash

WHO Global Alcohol Database

http://www3.who.int/whosis/menu.cfm?path=whosis,alcohol&language=english

ACKNOWLEDGEMENTS

Grateful acknowledgement is made to the following sources for permission to reproduce material in this book.

Figures

Figure 1.1a: © Held Collection/The Bridgeman Art Library, Figure 1.1b: Musée National du Moyen Age et des Thermes de Cluny, Paris, Lauros/Giraudon/The Bridgman Art Library; Figures 1.2 and 1.14: Lesley Smart; Figure 1.3: World Health Organization, *Adult Capital Consumption 2000*, www.euro.who.int/document/mediacentrefs1005e.pdf, The World Health Organization; Figure 1.4: World Health Organization (2005) *Differences in Drinking Patterns in the World*, www.euro.who.int/document/mediacentre/fs1005e.pdf, The World Health Organization; Figure 1.5: Adapted from Boreham, R. (2002), *Smoking, Drinking and Drug Use among Young People in England in 2002*, Class Licence Number C01W0000065 with the permission of the Controller of HMSO and the Queen's Printer for Scotland; Figure 1.6: Institute of Alcohol Studies (2007), *Binge Drinking*, Institute of Alcohol Studies, www.ias.org.uk; Figure 1.7: Mental Health Foundation (2007), *Why Do People Drink Alcohol?*, Mental Health Foundation; Figure 1.8: Matt Cardy/Getty Images; Figure 1.11: Leon, D. A. and McCambridge, J. (2006), 'Liver cirrhosis mortality rates in Britain from 1950 to 2002: an analysis of routine data', *The Lancet*, London, Elsevier Science Limited; Figure 1.12: Bryan Rosengrant/Flickr Photo Sharing; Figure 1.13: The Cabinet Office (2003), *Alcohol Misuse: How much does it cost?*, Class Licence Number C01W0000065 with the permission of the Controller of HMSO and the Queen's Printer for Scotland;

Figure 2.2:Accelrys; Figure 2.3a: Guy Grant; Figures 2.7a, 2.11 and 2.12 :Lesley Smart; Figure 2.7b: Scott Robinson/Flickr Photo Sharing; Figure 2.8: © Jack Sullivan/Alamy; Figure 2.9: Matt Riggott/Flickr Photo Sharing; Figure 2.10: Biological Sciences Curriculum Study, science.education.nih/gov/supplements/nih3/alcohol/guide/info: alcohol.htm, Biological Sciences Curriculum Study (BSCS); Figure 2.13: Image courtesy of Lifescan; Figure 2.14a: Internet Encyclopaedia; Figure 2.14b: Joel Creed/Ecoscene;

Figure 3.4a: Eye of Science/Science Photo Library; Figure 3.4b: Dr R. Dourmashkin/Science Photo Library;

Figure 4.2: www.CartoonStock.com; Figure 4.3: Dawn Partner; Figure 4.4: Courtesy of Dr Kaye H. Kilburn, Neuro Test Inc; Figure 4.12: Carol Midgley;

Figure 5.1: Paton, A. and Touquet, R. (2005), 'Risks associated with concentrations of alcohol in the blood', *ABC of Alcohol*, Oxford, Blackwell Publishing Limited; Figure 5.2: www.CartoonStock.com; Figure 5.4: Courtesy of Hilary MacQueen; Figure 5.5: Arthur Glauberman/Science Photo Library; Figure 5.6: Daniel Hommer M.D., National Institute on Alcohol Abuse and Alcoholism; Figures 5.7 and 5.8: Wattendorf, D. J. and Muenke, M. (2005) 'Fetal alcohol spectrum disorders', *American Family Physician*, Academy of Family Physicians;

Figure 6.1: Eye of Science/Science Photo Library; Figure 6.2: Assunta Del Buono/John Birdsall Photography.

Tables

Table 4.1: American Psychiatric Association (1994) *Diagnostic and Statistical Manual of Mental Disorders*, 4th edn, American Psychiatric Association.

Every effort has been made to contact copyright holders. If any have been inadvertently overlooked the publishers will be pleased to make the necessary arrangements at the first opportunity.

INDEX

Entries and page numbers in **bold type** refer to key words which are printed in **bold** in the text. Indexed information on pages indicated by *italics* is carried mainly or wholly in a figure or a table.

APPENDIX

Examples of prefixes which are commonly used with units of length, mass and volume*

prefix	prefix abbrevia-tion	prefix meaning	prefix as a power of ten	prefix as a number or fraction	examples of:		
					length	mass	volume
giga	G	times one billion	10^9	1 000 000 000			
mega	M	times one million	10^6	1 000 000			
kilo	k	times one thousand	10^3	1000	kilometre km	kilogram kg	
					metre m	gram g	litre l
centi	c	one-hundredth	10^{-2}	$\frac{1}{100}$	centimetre cm		centilitre cl
milli	m	one-thousandth	10^{-3}	$\frac{1}{1000}$	millimetre mm	milligram mg	millilitre ml
micro	μ	one-millionth	10^{-6}	$\frac{1}{1\,000\,000}$	micrometre μm	microgram μg	
nano	n	one-billionth	10^{-9}	$\frac{1}{1\,000\,000\,000}$	nanometre nm		
pico	p	one-trillionth	10^{-12}	$\frac{1}{1\,000\,000\,000\,000}$	picometre pm		

*Gaps in the table simply mean that the term is not commonly used.

Very large and very small numbers are conveniently expressed as a *power of ten*. For instance 1000 equals $10 \times 10 \times 10$, and this is written 10^3 (ten-cubed, or ten-to-the-three). At the other end of the scale $\frac{1}{1\,000\,000}$ equals $\frac{1}{10\times10\times10\times10\times10\times10}$ and is written as 10^{-6} (ten-to-the-minus-six).